SEX,
BOMBS,
AND
BURGERS

SEX, BOMBS, AND BURGERS

How War, Pornography, and Fast Food Have Shaped Modern Technology

Peter Nowak

LYONS PRESS
Guilford, Connecticut

An imprint of Globe Pequot Press

For Ian, Greg,
Richard, and Charlie

To buy books in quantity for corporate use
or incentives, call **(800) 962–0973**
or e-mail **premiums@GlobePequot.com.**

Library of Congress Cataloging-in-Publication Data is available on file.

ISBN 978-0-7627-7274-2

Printed in the United States of America

10 9 8 7 6 5 4 3 2 1

The author's views and opinions are not necessarily shared by the interview subjects or
their organizations and companies.

ᴣ314

CONTENTS

INTRODUCTION

A Shameful Trinity

Researchers believe that early human males developed the ability to walk on two feet in order to free up their hands for carrying food. It was the only way for some to get sex without fighting for it.

We are living now, not in the delicious intoxication induced by the early successes of science, but in a rather grisly morning-after, when it has become apparent that what triumphant science has done hitherto is to improve the means for achieving unimproved or actually deteriorated ends.[1]

—ALDOUS HUXLEY

The inspiration for this book came from the unlikeliest of sources: Paris Hilton. I wish it was some deeper or more sophisticated source, like the many scientific journals I've read, a PBS documentary I'd seen or even *Wired* magazine, but nope. My muse, I'm ashamed to admit, was a hotel heiress with no discernible talents.

It was 2004, at the very beginning of the young blonde's meteoric rise to celebrity. The Internet was aflutter with a video of Hilton, then twenty-three, having sex with her boyfriend, fellow socialite Rick Salomon. There was, as is usually the case with celebrity sex tapes, a debate over whether the video had been purposely leaked to raise Hilton's public profile. Regardless, it certainly succeeded in getting attention. The video intrigued me, not because of the sex or the celebrity-to-be, but because a good portion of it was green. The naked flesh on display was not a rosy pink, but rather monochromatic hues of emerald. This was, I realized, because the video had been shot in the dark using the camera's night-vision mode. While most viewers marveled at Paris's, er, skills, I was interested in the technology being used behind the scenes. Welcome to the life of a nerd.

As a technology journalist, I'm used to wondering what's under the hood, so to speak, and thinking about such cultural events in ways the non-technically minded, thankfully, never consider. When CNN trotted out the world's first televised "holograms" during the 2008 American presidential election and compared them to R2-D2's projection of Princess Leia in the first *Star Wars* movie, alarm bells rang and led me to discover that they were in fact "tomograms"—three-dimensional images beamed onto the viewer's screen and not into the thin air of CNN's studio. Similarly, most people enjoy Lego toys for their simplicity. Me? I couldn't help but wonder how designers decided on the optimum number, shape and variety of pieces in each set. So I called them to find out. It turns out that there is a lucky group of Lego employees who test-builds sets, using three-dimensional modeling software to create new pieces as they are needed. The software also prices the sets based on the number of parts, so designers can add or subtract pieces to get the kit to their target cost.

Such are my nerdy preoccupations; these are the stories I write in my daily life as a journalist.

I knew I had seen Paris Hilton's night-vision technology before. The notion nagged at me for days before it finally hit: the first Gulf War, also known as Operation Desert Storm. More than a decade earlier, a coalition of countries led by the United States had gone to war to liberate Kuwait from a brash takeover by Saddam Hussein's Iraq. I was too young for the televised reports of the Vietnam War so Desert Storm was the first big military action I had seen, played out on CNN as it was. The images that defined the war for me were the nighttime bombing raids—the barrage of anti-aircraft fire arcing upward, followed by huge explosions on the ground. Like the sex video, the most memorable images of Iraq's defeat were, for me, bathed in green.

It got me wondering what other consumer technologies are derived from the military. The more I delved into it, the more I found that just about everything is. From plastic bags and hairspray to vitamins and Google Earth, military money has funded the development of most of the modern items we use today. I also found many other links between war and the technology used in pornography, which is basically what Paris's video was. The porn industry has been quick to adopt every communications medium developed by the military, from smaller film cameras to magnetic recording (which led to VCRs) to lasers (which led to DVDs) to the Internet. Porn companies jumped on these technologies well before other commercial industries, thereby providing the money needed to develop them further.

The technological savvy of these two industries should come as no surprise. Lust and the need to fight or compete are two of the most primitive and powerful human instincts. They are our basest needs, a duo of forces that drive many of our key actions. Despite centuries of trying to deny, avoid, cure or otherwise suppress these forces, we have so far failed to find any course of action other than satiating them. As a result, catering to these needs has become big business. And big business needs technology to stay current and competitive.

Of course, there is another powerful urge that drives us: the rumbling in our bellies. At about the same time as Paris was getting famous, I was just starting to read the labels on grocery store shelves. Like anyone entering that phase of life where the metabolism starts to slow down—the tardy thirties, as I like to think of them—I was actually starting to care about what

I ate and therefore becoming concerned about the amount of glucose, fructose, phosphoric acid, sodium hydrogen carbonate and other assorted chemicals I was putting into my body. If you've ever read those labels and come across ingredients you can't pronounce, you've probably realized—as I did—just how much technology goes into our food.

As eye-opening as this was, though, it really shouldn't have come as a surprise. Our need for food is the most elemental instinct of them all, trumping all others, because without food we simply can't survive. It's understandable then that throughout history, we've used every resource at our disposal to ensure we have enough food on hand. Food has always been linked to power, and thereby to conflict. Historically, he who has had the most food has typically had the most power. And the best way to create lots of food is through technology. Ultimately, the more technology you have, the more food you have and the more powerful you are. This doesn't just apply at a macro level, either—in any society, a wealthy individual is a well-fed individual.

Our war-, sex- and food-related instincts go well beyond technology—they influenced human evolution itself. A recently unearthed hominid skeleton—4.4 million years old, the oldest discovered thus far—has presented evidence that war, sex and food were the three factors that led to humans getting up off all fours to become bipedal. Researchers at Kent State University in Ohio believe that early human males competed for female attentions by fighting it out. As with most apes, the ones who ended up with a mate were always the strongest and fiercest. Lesser males, however, also succeeded in getting female attention, but they used a different tactic—they brought them gifts. At the dawn of humanity, there was of course only one gift that mattered: food. Researchers have postulated that these lesser males had to learn how to walk on two feet in order to free up their hands so that they could carry this food to the females.

Millions of years later, little has changed. People still fight for food and sex, and we still use food (and other gifts) to try to get sex. These hard-wired, intersecting instincts have, over time, become our obsessions. Open any newspaper or watch any television broadcast and you'll see the proof. Endless broadcast hours and column inches are given over to the latest updates on the wars in Iraq, Afghanistan and elsewhere, the ongoing obesity epidemic or the latest diet craze, and the sex lives of celebrities or politicians caught in prostitution scandals. War, food and sex are everywhere because we demand them.

We feel compelled to fight each other, to compete and amass more than our neighbors have, whether through physical combat, political battle, verbal sparring or even just sports. War is an integral part of the human experience. Lust, meanwhile, leads people to do stupid, stupid things, from risking unwanted pregnancy and diseases by having unprotected sex to courting identity theft by giving their credit card numbers to shady websites or provoking the loss of their families and relationships by conducting poorly concealed affairs. Mythology is rife with conflicts fought over sexual jealousy, such as the Trojan War, which started when the King of Troy's son stole the King of Sparta's wife, while even Adam disobeyed God by eating the forbidden fruit in hopes that Eve would get it on. The need for food is just as basic. Wars still start over the land that produces food, while in the most extreme cases, a lack of food even drives man to eat fellow man.

Huge industries have developed around each need; war, sex and food are not only humanity's oldest businesses, they are some of the biggest as well. But the question does arise: with three such basic instincts, why the need to tinker? If our needs are so elemental, why does meeting them require such ongoing innovation? The reasons, it turns out, are many.

The Whys Have It

As Thomas Friedman explains in *The World Is Flat*, the iron law of American politics is that the party that harnesses the latest technology dominates. FDR did it with his fireside radio chats, JFK did it with his televised debates against Nixon and the Republicans did it with talk radio. More recently, Barack Obama did it through his use of social media such as YouTube and Twitter, leaving his opponent John McCain—who admitted he didn't even know how to use his computer—looking like an out-of-touch Luddite.

The American government believes that this rule of politics, a diplomatic form of war, extends to the actual battlefield, which is why its military is both a big creator of technology and a long-time early adopter. Most obviously, superior technology gives an army an advantage over its enemies, at least temporarily and especially psychologically. Once the United States dropped the atomic bomb on Japan, there was no question the Second World War was over. Similarly, the Gulf War was quick and decisive—it lasted only two months—because coalition technology such as smart bombs, GPS and night vision allowed for a devastating aerial bombardment. By the time the ground war started, Iraqi troops were ready to give up. Today, insurgents in Iraq are more determined and aren't as intimidated by technology—still, imagine how scary it must be to see robots firing at you. That's the sort of edge the U.S. military is always looking for, says Colonel James Braden, who fought in Desert Storm and is now a project manager for the marines. "I don't ever want to be in a fair fight," he says.[2]

As military officials are quick to point out, technology also saves lives. Whether it takes the form of body armor, night vision or robot scouts, that's worth any price. "America, Canada, the U.K., Australia and others are countries willing to make a substantial investment in the safety of their sons and daughters, and thankfully so," says Joe Dyer, a retired navy vice-admiral and executive with robot maker iRobot. "The value we place on any life is the engine that drives that. Two things—a reduction in the number of people that we put in harm's way and an improvement in the survivability for those that go into battle—those are the central pillars of American technology."[3]

In recent years, technology has focused on making fighting not only safer, but more comfortable. Pilots are now flying robot drones in Iraq from air bases in Nevada, then going home to pick up their kids from soccer practice. Front-line troops, meanwhile, get to spend their downtime with Xbox video games. There's no more playing the harmonica or writing long letters to Betty Sue back home. Now they've got email and the Internet.

Besides saving troops' lives and making them more comfortable, technology can limit the damage inflicted on the enemy, particularly on civilians. This is important in weakening the pervasive circle of hate, where the ruined lives of one war become the revenge-bent soldiers of the next. "There's an efficiency in attack, but there's also a minimization of damaged infrastructure and collateral damage and lives lost in military forces and civilians," says Dyer. Today's enemies, he adds, may become tomorrow's allies.

In the United States, investment in military technology has also become a pillar of scientific research. The military and its various labs, especially the Defense Advanced Research Projects Agency, do much of the long-term work that is too expensive or far-out for industry. The Internet is one example; created by DARPA during the Cold War, the communications network took a full twenty years to get up and running. Vint Cerf, the computer scientist who made the network's first connections, says civilian companies don't have the patience for such projects: "If I had to cite one aspect of DARPA's style, it was its ability to sustain research for long periods of time."[4] As we'll see, since its inception in the late fifties, DARPA has generated one important technology after another, including cellphones, computer graphics, weather satellites, fuel cells, lasers, the rockets that took man to the moon, robots and, soon, universal translation, thought-controlled prosthetics and invisibility. None of these was developed in one fiscal quarter. War historians and social scientists say DARPA has "shaped the world we live in more than any other government agency, business or organization."[5]

National labs staffed with rank-and-file scientists are also key. For these institutions, performing research for the military is a win-win-win situation: they get funding to do basic scientific work and the military gets the results, which are ultimately passed on to the civilian world. Lawrence Livermore National Laboratory, east of San Francisco, is a good example. The lab was opened in 1952 with a mandate to conduct defense-related research. Livermore is responsible for looking after the U.S. nuclear arsenal, so much of its research has concentrated on that area. You'd think such a lab would operate under the strictest of security—and it does, but the scientists there are only too happy to talk about how their weapons work has spun off into everyday life. Those spinoffs have been numerous: three-dimensional collision-modeling software, originally devised to predict the impact of bombs, now used by carmakers to simulate crashes and by beer companies to test new cans; genetic research that kicked off the Human Genome Project; laser-hardening systems that are now helping planes fly farther; and the latest, a proton accelerator beam that promises to revolutionize cancer therapy. George Caporaso, a scientist who has worked on beam research at Livermore for more than thirty years, says there's no way such breakthroughs could have happened in the civilian world. "When I came here I saw the scale of things that were being done, things private industry wouldn't dare take a risk to explore, only the labs could do that," he says. "Really high-risk, high-reward things . . . very important problems for national defense and security . . . things that in many cases no one else can do."[6]

A good deal of technology thereby comes from military spending. And with developments tied to how much money goes into research, the pace of innovation looks set to increase. In 2009 the world's combined militaries spent an astonishing $1.5 trillion, or about 2.4 percent of global gross domestic product. That amount, up 49 percent since 2000, was a new record. For the most part, this is an American story, as nearly two-thirds of the increase came from the United States.[7] American military spending increased 7.7 percent in 2009 to $661 billion, or 54 percent of the global total, more than six times more than second-placed China, with a relatively paltry $100 billion.[8]

The United States spends so much on its military that the Pentagon's secret "black budget" of $50 billion is more than the entire defense budget of most countries, including the United Kingdom, France and Japan, and more than triple Canada's.[9] Much of this money flows to DARPA, universities and national labs such as Livermore. By one estimate, as many as a third of major university research faculty have been supported by national security agencies since 1945.[10] No wonder the United States produces a disproportionate amount of the world's science and technology: despite having only 4 percent of the world's population, the country spends about half the world's research and development dollars.[11] It has also made technology part of the cultural fabric. As a U.S. Army report states, "Technology is part of how Americans see themselves, to reach for it is instinctive."[12]

Some military inventions have been commercialized in obvious ways; the atomic bomb, for one, was turned into nuclear energy, while jet fighters became jetliners. This book will feature many less obvious spinoffs, like how radar led to the microwave oven or spy satellites spawned Google Earth. The important thing is that the technologies eventually came to market. "The whole history of Silicon Valley is tied up pretty closely with the military. Integrated circuits were designed to guide warheads. All the constraints that were necessary to make that successful drove a lot of the miniaturization to work," says John Hanke, who helped create Google Earth. The military is "willing to pay millions of dollars per user to make it possible. Things have this very high value that you don't necessarily see in the consumer space. Once the technology and the basic R&D is paid for, then companies start looking for those secondary markets where you can take the things they know how to do."[13]

We've come to the point where it's almost impossible to separate any American-created technology from the American military. Chances are, the military has had a hand in it and industry has been a willing partner. In the case of Livermore's collision-modeling software, car companies have been working with the lab since the 1980s to refine the tool. The carmakers and the lab swap software code back and forth in an effort to make it better. "There's give and go in both directions," says Ed Zywicz, one of Livermore's programmers.[14] James Braden, the Marine colonel, says the military's relationship with industry has never been as good as it is today. "As you improve a ground combat vehicle, some of that spins off into the automotive industry and on the flip side of that, many of the things the automotive industry comes up with spin off into our ground combat vehicles," he says.

War also drives economic activity, a truth that governments have always known. The Second World War, for example, ended the Great Depression by stimulating demand for everything from shoes to steel to submarines. In 1933 a quarter of the American workforce was unemployed and the stock market had lost 90 percent of its value since the crash of four years earlier.[15] President Franklin D. Roosevelt's New Deal, a package of social reforms, steered the economy in the right direction, but the crisis was still in full flow when war broke out. The United States saw steady economic recovery, first as it supplied its allies and then when it formally joined the war in 1941. The entire might of American industry moved to support the war effort and the nation reaped the benefits. In 1944, when defense spending reached an astounding 86 percent of the federal budget, gross national product climbed a correspondingly huge 28 percent.

In dollar terms, GNP went from $88.6 billion in 1939 to $135 billion in 1944. These are growth rates that economists and stock investors see only in their dreams.

The benefits went beyond mere numbers. By 1944 unemployment had dropped to 1.2 percent of the population—it has never again been that low—and even those people who couldn't normally find jobs, including many women and blacks, were gainfully employed. The contributions of those particular demographics during the war also did much to further their respective rights movements in the years that followed.

History looks to be repeating itself since the "war on terror" began in 2001. American defense spending has shot up 74 percent since 2002. Gross domestic product saw a correspondingly solid rise of 2.9 percent on average between 2001 and 2008, before the global recession set in. In 2007 unemployment was among the lowest it has ever been. The United Kingdom, a major ally in the war on terror, experienced similar benefits. GDP grew an average 2.3 percent while unemployment hit its lowest point in more than two decades in 2007. This growth was not coincidental. As of early 2011, the war on terror had cost the United States more than $1.1 trillion—but much of that taxpayer money didn't just evaporate, it went right back into the industries that did the heavy lifting.[16]

Trying Out New Positions

The pornography industry spends money on technology too, though it generally doesn't create it: there are no labs with scientists in white coats working on better porn innovations.

Instead, the industry exerts its influence as an early adopter. Porn dollars often make it possible for technology-creating companies to stay afloat and improve their innovations to the point where they're ready for a mainstream audience. Just ask Brad Casemore, the product manager for a small Toronto-based company called Spatial View. In 2009 the company introduced software that allows iPhone owners to view 3-D photos and videos. A porn company, Pink Visual, was one of the first to license the technology and produce content with it. "The people in the industry feel a great sense of urgency to stay ahead of their competitors and, no pun intended, they have a really demanding audience," Casemore says. "It's an audience that's continually looking for new ways to access content."[17]

Having been at the forefront of the industry since its first issue in 1974, *Hustler* has seen more technological change than just about anyone in porn. Michael Klein, president of Larry Flynt Publications (which publishes *Hustler*), agrees with Casemore's assessment. Porn companies have to be on top of the latest technology because they are in an extremely competitive industry. If they're too slow in getting their content to people, somebody else will beat them to it. "You need to be out there and get the product out as quickly as you can to consumers," he says.[18]

Porn companies are able to adopt new technologies quickly because they are usually small and, like the actors they hire, quite flexible. Tera Patrick, one of the industry's biggest stars, started her own company, Teravision, in 2003, with then-husband Evan Seinfeld, bassist and singer with the heavy metal–hardcore band Biohazard. They say porn companies can experiment with technology because they don't have to go through layers of management. "The mainstream is slow on the go. They need a hundred people to make a decision. In a corporate structure, people are afraid to go out on a limb because if something goes wrong, they could lose their jobs," Seinfeld says. "In adult entertainment, if you came to Tera and myself and said, 'Hey, I've got this new technology to deliver your content to people via an iPhone,' we could run a check on your company in a matter of days, have a contract and be up and running within a week." Patrick agrees: "If somebody pitches me something, we can move at light speed. In some cases the early bird does get the worm."[19]

Even the bigger adult companies like LFP and Playboy Enterprises don't feel hemmed in by their corporate responsibilities and behave like entrepreneurs. If a mainstream movie studio wants to try something new, Klein says, "they have to go back and get Tom Cruise's permission or Tom Hanks's permission, the director's permission, the writer's permission. We secure all those rights and we control the movies we have when we do the deal so it's easier for us to try all these things." The result is that there are thousands of privately owned entrepreneurial porn companies with lots of money to burn. As porn star Stoya points out, they can afford to say, "Yeah, let's give it a shot—if it doesn't turn out, it's no big deal."[20]

This flexibility attracts non-traditional entrepreneurs who have trouble fitting into the mainstream. Scott Coffman, a serial entrepreneur who had tried to create, among other things, board games and herbal supplements, saw potential in video-on-demand because of the way people were pumping quarters into peep-show booths in adult video stores. He started the Adult Entertainment Broadcasting Network, a website where visitors can pay to watch porn on a per-minute basis, to try out the peep-show concept online. AEBN, Coffman says, now pulls in a hundred million dollars in revenue a year, making it the largest video-on-demand provider in the world, in or out of porn. "I thought that's the way we should be selling adult entertainment online, because guys only want to come and watch for a short amount of time. We saw the potential when most companies didn't really care about it," Coffman says. "We showed that you could make money on the Internet with video. We didn't just show the way in technology, but also in economics."[21]

After graduating from the University of Southern California's prestigious film school, Ali Joone started Digital Playground in 1993. In 1997 he bet the company on an emerging home-video technology, DVD, a move that mainstream producers were reluctant to make in light of such previous flops as laser disc and CD-ROM. "Everybody in the industry thought DVD was a fad, mainly because CD-ROM came and went. I never looked at CD-ROM as the final product because the image was small. But when you looked at DVD, the image quality was great and it was a big leap from VHS," he says. "The mainstream waits until there's a number of players out there."[22] Porn is also a medium that allows for unparalleled freedom and creativity, Joone says. "Adult is the last place where you can do independent filmmaking. You can make any movie you want as long as it has sex in it. As a creative person, your boundaries are huge."

During the course of researching this book, I was surprised to learn that Paul Benoit, a former high-school friend, was the chief operating officer for the company that produces the popular porn site Twistys. I never pictured him as a porn executive, but as he explains, he was attracted by the marketing opportunities and creative freedom. "What's always fascinated me about the industry isn't the breasts or the sex, but how willing it is to push the envelope," he says. "Being in the industry means doing whatever our imaginations can generate."[23]

Porn companies are forced to innovate, Benoit says, because they are one of the most stigmatized, marginalized and even hated industries there is. As such, they face limitations, regulations and prosecution at every step. Credit card firms, for example, charge them larger transaction fees because they run a higher risk of customers wanting their money back (people don't tend to think transactions through when they're sexually aroused, which often results in buyer's regret). A growing number of countries are also looking into ways of outright blocking pornography online—and not just the violent and disturbing stuff, but simple nudity as well. One filter tested by the Chinese government even blocked pictures of pigs, which were mistaken for naked human flesh. And this isn't happening only in developing or devoutly religious countries like China and India, but in countries with supposedly strong free-speech laws such as Australia and the United Kingdom.

For porn companies, necessity really is the mother of invention. They have to get to new technologies quickly and exploit them for as long as they can, until regulators catch up. Then it's on to the next technology. "To deliver, you have to invent in order to circumvent other issues that may come up," Benoit says.

Decades of technological innovation and virtually unlimited demand have given the industry a veritable licence to print money. Many producers have resisted going public because of the red tape, including responsibilities to management and shareholders, that inevitably follows. As with traditional businesses, many believe that once shareholders enter the equation, the ability to push boundaries lessens. Non-transparent private companies are also better at hiding information, such as how much money they're generating, which helps them fly beneath the radar of authorities and tax collectors.

All of this explains why there are only a handful of publicly traded companies that deal in sexual content, including Chicago-based Playboy, Colorado-based New Frontier Media, Barcelona-based Private Media and Germany's Beate Uhse. Playboy and Beate Uhse are the biggest, each bringing in around $300 million in revenue a year. There are many big independently owned companies that reportedly pull in hundreds of millions in revenue each, including a pair of Japanese firms, Soft on Demand and Hotuku, London-based Dennis Publishing and a host of American companies, including Digital Playground, AEBN, Vivid and Wicked Pictures.[24] Aside from the big boys, there are thousands of smaller companies such as Twistys and Teravision. Accurate numbers are difficult to come by because of the industry's lack of transparency, and because, as technology historian Jonathan Coopersmith puts it, "everyone lies."[25]

The estimates that do exist are jaw-dropping. The worldwide porn market has been pegged at a whopping $97 billion, or more than the combined revenue of some of the biggest Internet companies, including Google, Apple, Amazon, eBay and Yahoo. China, South Korea and Japan are the top three porn revenue generators, although in the case of China the numbers are likely skewed toward the manufacture of goods like DVDs and sex toys rather than consumption, given the country's strict anti-pornography rules. The United States rates fourth with an estimated $13 billion spent, followed by Australia and the

United Kingdom with about $2 billion each. That worldwide revenue amounts to $3,075 being spent on porn *every second*, $89 of which is on the Internet.

While the United States only places a technical fourth in estimated revenue (though its $13 billion still exceeds the combined revenue of the three biggest television networks, ABC, CBS and NBC, and is close to the total intake of the NFL, NBA and MLB), porn innovation, like military innovation, is still very much an American story. American producers have led every medium shift, from film to videotape to DVD to the Internet, where 89 percent of adult websites originate in the United States. The rest of the world doesn't even come close; Germany is second with 4 percent of porn sites, followed by the United Kingdom with 3 percent. Asian countries make up a mere sliver of the total porn-site pie.[26] While pornography is a global market, the worldwide industry looks to American producers to find out what's next.

Keeping Up with the McDonalds

The motivation behind innovation in food technology is simple: as the world's biggest industry, the stakes in food are huge and the competition is fierce. Producers must cater to customers' tastes, even if they are unreasonable, which is more the case in the food industry than in perhaps any other. We want burgers and fries, but we also want them to be healthy. We want fresh produce year-round, even in the dead of winter. We want food immediately because time is our most precious commodity. We also want our food cheap so that we can spend our money on "more important" things, like new cars and flat-screen TVs. All of this requires technological innovation. For producers, catering to these needs is not easy, but the smallest changes can provide big rewards. Huge differences in profit can often be measured in seconds. If a fast-food restaurant can shave a few seconds off serving a burger and then sell millions of burgers, the technology that will help it do that is well worth the investment.

Historically, stabilizing a population's food supply has meant improving agricultural methods, harvesting tools, equipment, transportation and infrastructure, all of which inevitably involve technology. The Green Revolution, which swept the world in the fifties and sixties, applied agricultural innovations such as hybrid seeds, chemical fertilizers, pesticides and irrigation methods. Norman Borlaug, the American scientist known as the father of the Green Revolution, was fond of pointing out how integral those technologies were to prosperity. "Civilization as it is known today could not have evolved, nor can it survive, without an adequate food supply," he said during his Nobel Peace Prize acceptance speech in 1970.

After stability, the impetus for innovation becomes portability and export, and it's here that the military ties have typically come in. During the Second World War, the American government spurred massive investment in new preservation techniques such as spray-drying and dehydration. These new technologies were needed to make foods last longer and withstand the rigors of traveling thousands of miles across the ocean, to be consumed by soldiers. The investments paid off, resulting in what food historians have called the best-fed troops in history. While the average American male civilian ate 125 pounds of meat in 1942, a typical soldier was allotted 360 pounds. Some troops were getting as much as eleven pounds of food per day while civilians made do with four.[27]

When the war ended, the United States emerged as a food power with a huge advantage over the rest of the world. By 1946 the country was established as a major exporter, producing 10 percent of the world's food.[28] Its position as the first country to achieve political, economic and agricultural stability was strengthened by the fact that Europe had just seen much of its infrastructure destroyed, and by coincidental large-scale crop failures in Asia. In 1946, while Americans wallowed in the luxury of too much food, more than 600,000 Chinese died of starvation and 125 million Europeans faced malnutrition.[29] Today, the United States is the largest exporter of food and livestock, with nearly double the output of the next country on the list, France, followed by Germany, the Netherlands, China, Spain, Belgium and Canada (it's no coincidence that the top three food exporters in the world are G8 countries).[30] The American food system has, in fact, created so much abundance that it wastes more food than many nations produce. Americans spend half a trillion dollars every year at supermarkets and another half a trillion at restaurants, half of that at fast-food outlets. They also end up throwing away half of the food that is ready for harvest.[31]

Once stability and exports were established, profit became the main motivator for continued innovation in the food industry. This led to consolidation into fewer and fewer major powerhouses over the second half of the twentieth century. These huge, publicly owned companies must earn profits for shareholders or risk being consolidated themselves. Many of them are based in the United States and several, including Pepsi, Kraft, Coca-Cola, Tyson Foods (the world's biggest chicken processor) and McDonald's, are Fortune 500 companies. Together, they comprise the largest industry in the world, dwarfing even the military, with annual revenue of about $4.8 trillion, or 10 percent of global economic output.[32] With that much money at stake, the industry is incredibly competitive— companies must continually come up with new ways to generate profits. There are two ways to do this: squeeze new operating efficiencies into the business or think up new products.

When it comes to improving efficiencies, the same rules apply as in boosting economic productivity. In economics, a worker's productivity is measured in units produced per hour worked. Economists agree that there are basically two ways to boost this output—either a business hires more employees to spread the workload or it invests in technology that allows each individual worker to do more. This rule also applies to agriculture, more specifically to farmland. The only way to boost agricultural output is to increase the amount of land being used or invest in more efficient technology. Since populations and cities are growing and arable land is decreasing proportionately, well ... there's really only one option. During the Green Revolution, this meant hybrid plants that made more efficient use of the land. But it hasn't been enough, so biotechnology companies such as Monsanto have turned to genetically modified plants and animals to boost efficiency once again. If you can grow a stalk of corn that produces more useful kernels and fewer useless leaves and stalks on the same amount of land, why not do so? Or so goes the rationale.

At the product level, processors are continually looking for new ways to cater to customers' wants, which have typically centered on food that's cheaper or takes less time to make. Recently, consumers have also started demanding healthier foodstuffs. Companies such as Hormel and Unilever, for example, are incorporating a processing method that uses highly pressurized water in conjunction with microwaves to precook meats and vegetables. This new method cooks much faster than the old steam-based system and the food retains more of its natural taste and nutrients. The process was originally developed for the military but is now going mainstream because it gives the companies a competitive edge—healthier foods—over rivals. "They put in a lot of effort because they see there's a benefit for their civilian market," says Patrick Dunne, a food scientist at the U.S. military's Natick food lab, which co-developed the technology with the companies.[33]

Every food company is looking to save money in production, which is why, to take just one example, you may have noticed a steady increase in the number of foods at the grocery store that now come packaged in pouches rather than cans. These "retort" or flexible pouches, again developed by the military, are lighter than cans, which cuts down on shipping costs.

During the second half of the twentieth century, many families saw both parents go off to work for the first time, so less time was available to prepare meals. Not only did this force companies to come up with quick and easy-to-prepare meals, it also spurred the rise of fast-food chains. Technology figured prominently in both. Through the power of processed foods and new cooking innovations such as the microwave oven, a meal that took one hour to produce in 1965 was shaved down to only thirty-five minutes by the mid-nineties.[34] Fast-food chains such as McDonald's built themselves on technology great and small—from big innovations such as frozen burgers and fries down to the humble ketchup-squirting gun—to provide inexpensive and easy meals for consumers in a hurry. All of this technology dramatically changed eating patterns—by 2007,

90 percent of Americans ate food produced outside of the home every day, much of it fast food.[35]

McDonald's, the largest restaurant chain in the world with more than 31,000 outlets and annual revenue exceeding $20 billion, is constantly searching for ways to improve food safety and quality and customer satisfaction, in order to ensure the return visits it relies on. The company recently introduced automatic fry dispenser baskets and automated drink-pouring machines to help cut seconds off order times. McDonald's has also established a Technology Leadership Board, where store managers contribute ideas for improving the system. Customers are hired to visit its test lab in Romeoville, a suburb of Chicago, to try out new inventions. "Our customers don't come in for technology, they come in to get great service and great food in a clean environment," says Dave Rogers, senior director of the restaurant solutions group for McDonald's Canada. "We look at it as an incredibly important enabler to run our restaurants more efficiently. We're always looking for ways to save seconds."[36] Given that McDonald's is the world's largest purchaser of beef, pork and potatoes and the second-biggest buyer of chicken (KFC is first), its technological advances have major impacts, not just on the rest of the fast-food industry but on the world's entire food-production system.

There is also one other reason to invest in food innovation: hunger is a major motivator for turning people toward war and terrorism. When you and your family are starving and the local army or al Qaeda offers you food, you probably won't ask too many questions. With the world's population set to double in the next fifty years and global warming expected to create food shortages for an additional 200 million to 600 million people in the developing world, the problem is only going to get worse.[37] Part of the solution will be to move food around better (including making use of the staggering amount being thrown out by Americans) and part of it will involve more efficient food production. Both will involve more technology, not less.

Side Effects

The technologies of war, sex and food haven't just changed the goods we buy, they have also changed us as people and as a society—sometimes for the better, sometimes for the worse—in ways we rarely consider. These issues are not new. Archeological evidence suggests that one of humanity's first inventions, fire, was used for the purposes of war, sex and food. Prehistoric man used fire to frighten enemies and animals. He also used it to illuminate caves so that he could paint, among other things, depictions of sexual acts on the walls. And cave men (and women) used fire to sterilize their food, perhaps as Fred Flintstone did in cooking his Brontosaurus burgers. Moving forward through time, ancient societies invented iron and used it to create weapons, while the development of shipbuilding allowed for the spread of empires. Gutenberg's first printed book may have been the Bible, but it was soon followed by erotically charged works such as The Decameron and The Canterbury Tales, which helped pay his bills. The invention of canning, meanwhile, helped Napoleon march his troops around Europe. Had the microwave oven been invented, he might even have succeeded in his attempt to conquer Russia.

The modern technological world began in the mid-twentieth century as global war erupted. The Second World War was an unprecedented conflagration involving more nations than any previous conflict. The stakes had never been so high—the

Axis powers were bent on genocidal world domination while the Allies were determined to prevent a future under Fascism. Both sides looked to technology to gain an advantage over their enemies, and both sides invested heavily. A host of world-changing technologies emerged from this deadly competition, from jet airplanes to computers to rockets that took us to the moon. Technology ultimately provided the exclamation point to the end of the war, with the atomic bomb explosions in Japan heralding a new age of future hopes—and fears. The Second World War set in motion our continual, modern-age quest for new and better technology, a turbo-charged sprint to a world that is more advanced than even the most imaginative of science-fiction writers envisioned a century ago. The effects of this epic war are around us everywhere today, so it's a good place to begin our exploration of the technology of sex, bombs and burgers.

WEAPONS OF MASS CONSUMPTION

The Radarange microwave oven promised cooking convenience to housewives of the 1950s.

The view from atop Coventry Cathedral differs from that of many of Europe's old towers. Rather than cobblestone streets and centuries-old edifices, Coventry's oldest structure is surrounded by shopping malls. Off in one direction you can see a soccer stadium, in another a big blue IKEA store. There are few tourists scrambling for photos of landmarks here, just busy locals hurrying to get some shopping done during their lunch breaks as police cars zip by on the paved roads, sirens wailing. Radiating out from the cathedral in all directions are the hallmarks of twentieth-century construction: steel, concrete and glass buildings, both low- and high-rise, housing businesses and residences. There is little trace of antiquity in this city of more than 300,000 inhabitants, smack dab in the middle of England. Coventry resembles the new, manufactured metropolises of North America more than it does the Old World. But that wasn't always the case.

Coventry has had three cathedrals in its history. The earliest, dedicated to St. Mary, was established in 1043 as a Benedictine community by Leofric, the Earl of Mercia, and his wife, Lady Godiva—the same woman who, according to legend, rode a horse naked through the streets to protest the excessive taxes her husband had imposed on residents. A statue commemorating her ride now stands in the middle of a pedestrianized outdoor shopping mall, just a stone's throw from the current cathedral.

Over the next few centuries the settlement around the church grew, mainly on the strength of its textile trade, and by 1300 Coventry was England's fourth-largest city. During the following two centuries, it became the kingdom's temporary capital on several occasions when the monarchy relocated to avoid rebellions in London. St. Mary's was also replaced by a grand Gothic church, St. Michael's, but this fell into disrepair during the sixteenth century when Britain's monasteries were dissolved. By the early twentieth century, Coventry had evolved into a major manufacturing center, particularly for cars—the city was the Detroit of the United Kingdom, headquarters for heavyweights Jaguar and Rover—and the population had risen to a quarter of a million. St. Michael's church, meanwhile, was elevated in status in 1918 to become the city's second cathedral. With the outbreak of the Second World War, Coventry, touted by its government as the best preserved medieval city in England, also became one of the country's top producers of airplanes and tanks, a status that made it a prime target for the Nazis.[1]

On the evening of November 14, 1940, German bombers commenced Operation Moonlight Sonata, Hitler's most ambitious and vicious attack on England. Luftwaffe bombers pounded Coventry with wave after wave of high explosives and incendiary bombs from dusk till dawn, killing more than 550 civilians, injuring thousands, destroying more than 4,300 homes and damaging three-quarters of the city's factories.[2] St. Michael's Cathedral, the city's figurative and historical heart, was destroyed, save for the miraculous sparing of its spire. The German attack, intended to hurt England's production capability and soften it up for an all-out invasion, was called "one of the worst bombardments from the air since the Wright brothers presented wings to mankind."[3] Hermann Göring, the Luftwaffe commander, boasted of the destruction and coined a new word to mark the occasion: *Koventrieren*, to "Coventrate" or raze to the ground.

In an editorial decrying the bombing, the *New York Times* pointed out that the horror of the Blitz had only happened because there was no defense against such a night-time assault. Anti-aircraft guns and mines strapped to balloons, a tactic seemingly borrowed from a Road Runner cartoon, brought down only a handful of attacking planes, which meant that "other great industrial centers and ports in England are exposed to the same fate whenever the moon is bright and the weather favorable to raiders." Until a new defense could be developed, Prime Minister Winston Churchill's warnings that "death and sorrow will be our companions on the journey, hardship our garment, constancy and valour our only shield" would continue to ring true.[4]

The development of such a defence was secretly underway in the ruins of Birmingham, which had been similarly "Coventrated." Physicists John Randall and Harry Boot were experimenting with an improved version of the cavity magnetron, a copper tube that generated microwaves. At the magnetron's center was a cathode that pumped out electrons, which were spun around the tube by an attached electromagnet, a process that gave the device its name. The electrons spun past cavities drilled into the tube and produced radio waves. Those waves were then emitted and, if they hit something, bounced back to their source. This echo effect let the device operator know that something had been detected, and pinpointed the object's position. Earlier versions of the magnetron, developed in the United States and Germany, were of limited use because they didn't generate much power and had poor range. Randall and Boot boosted the tube's power output a hundred-fold by drilling eight cavities into the magnetron instead of the standard two, and by adding a liquid cooling system. The result was a more powerful magnetron that was compact enough to be fitted into aircraft. The British government believed that giving planes the ability to see enemies at night would be a major advantage, perhaps enough to turn the tide of the war. The problem, however, was that Britain was cut off from its traditional European allies, now all under Hitler's thumb, and lacked the production capacity and manpower to produce the device in the large numbers needed.

Bring in the Yanks

Britain turned to its long-time pal, the United States, for help. With the Nazis pressing and time running out, Henry Tizard, head of British aeronautical research, set out on a voyage across the Atlantic in late September 1940, taking with him the nation's most valuable technological secrets, including blueprints and diagrams for explosives, rockets, self-sealing fuel tanks, the beginnings of plans for an atomic bomb and, the main attraction, the magnetron. Tizard put the magnetron's fate into the hands of Vannevar Bush (no relation to the presidents), an inventor, electrical engineer, entrepreneur and patriot who resembled a beardless Uncle Sam.[5]

In his early adulthood in the 1910s, Bush had supplemented his undergraduate studies at Tufts College near Boston by working at General Electric and as research director for American Radio, a small company started by his fellow students Charles Smith and Al Spencer. (The company achieved some minor success during the First World War with Smith's invention of the S-Tube, which eliminated the need to use batteries in radios, but was all but wiped out by the Great Depression.) In 1917 Bush received his doctorate in electrical engineering from Harvard and the Massachusetts Institute of Technology (MIT), and by 1923 had become a professor at the latter. In 1922 Bush and fellow Tufts engineering student Laurence Marshall teamed up with Smith and set up the American Appliance Company to market another of Smith's inventions, a refrigerator with no moving parts—its solid state making it less prone to breaking— but failed miserably when they found no takers. The trio's backup plan was an improved version of the S-Tube. They brought Al Spencer back on board, along with his younger brother Percy, and by 1925 American Appliance was earning a profit.[6] To avoid problems with a similarly named company operating out of the Midwest, the group renamed the business Raytheon Manufacturing—adding "theon," Greek for "of the gods," to the rays of light their tubes produced. For the beleaguered British, Raytheon proved to be a godsend indeed.

In his public service life, Bush had helped develop a submarine detector for the American government during the First World War, but the system was never used because of bureaucratic confusion between industry and the military. "That experience forced into my mind pretty solidly the complete lack of proper liaison between the military and the civilian in the development of weapons in time of war, and what that lack meant," he later recalled.[7] In 1932 Bush became vice-president and dean of engineering at MIT, then moved to the prestigious Carnegie Institute of Washington as president in 1939, to be closer to the corridors of government power. Lack of cooperation was something he would not tolerate during the new conflict. Along with a group of fellow science administrators, including MIT president Karl Compton, Bush pitched President Franklin D. Roosevelt on an organization that would oversee research and development work between industry and the military. Bush showed Roosevelt a letter that proposed his National Defense Research Council and the president approved it on the spot. "The whole audience lasted less than ten minutes . . . I came out with my 'OK–FDR' and all the wheels began to turn," he later wrote.[8] On June 12, 1940, the American military-industrial complex was born, with the patriot Vannevar Bush as its beaming father.

Bush was the chairman of the new NDRC while Compton was put in charge of developing radar. The first few meetings between Tizard's delegation and the new American military-industrial brain trust were cautious, like a high-stakes poker game with neither side wanting to reveal its hand. Compton cautiously showed the British visitors the low-powered magnetrons developed by American scientists, which thawed the atmosphere between the two camps. After seeing that the two nations were on the same path, Tizard proudly demonstrated the high-powered British magnetron to the astonishment of his hosts, prompting the envious Compton to order the immediate establishment of the Radiation Laboratory at MIT to develop the device further. Large electronics manufacturers including General Electric, Bell Telephone and Western Electric were brought in to mass-produce the magnetron, but they encountered a problem: because the gizmo had to be machine-chiseled from a solid copper block, producing it in mass quantities was difficult, time-consuming and expensive.

Both Compton and Bush, who by now had extricated himself from the day-to-day operations of Raytheon but still held a seat on its board of directors, were well acquainted with the company's talented lead inventor, Percy Spencer. Raytheon was small compared with the likes of GE and Bell, but the company was just down the road from MIT in Waltham, Massachusetts, so Spencer was called in to take a look at the magnetron.[9]

Percy Spencer was an orphan and, as a child, poor as dirt. His father died when he was eighteen months old and his mother abandoned him soon after, leaving him to be raised by his aunt and uncle in Howland, Maine. More bad luck struck at the age of seven when his uncle died. Spencer spent his childhood doing country chores such as saddling horses and chopping wood, and was so poor he used to hunt to eat. From the age of twelve he worked at a spool mill, starting before dawn and continuing on until after sunset.

The enterprising youngster was extraordinarily curious, though, and when it came time to install electricity in the mill, he volunteered to do it. He learned by trial and error and emerged from the project a competent electrician. When the *Titanic* sank in 1912, his imagination was sparked by the heroism of the radio operators who had helped rescue survivors. So he joined the navy and learned wireless telegraphy: "I just got hold of textbooks and taught myself while I was standing watch at night," he later recalled.[10] His self-education went so well that the navy made him head of wireless production during the First World War. By 1940 the Raytheon engineer was renowned among scientists at MIT. "Spencer became one of the best tube designers in the world; he could make a working tube out of a sardine can," one said.[11]

This reputation served Spencer well when he asked if he could take the magnetron, Britain's most closely guarded technological secret, home for the weekend. It was like asking the Queen if he could borrow the Crown Jewels. But with the combined brain trust of MIT vouching for Spencer, Henry Tizard reluctantly gave his blessing. Spencer returned with what now seems like a no-brainer of a suggestion: rather than carving the magnetron out of a single lump of copper, why not create it piecemeal from several sections?

Western Electric had already been awarded a $30 million contract to manufacture the magnetron tubes, but was only managing to produce about fifteen a day using the machining method. Spencer promised he could outdo that production with his alternative procedure, so MIT gave Raytheon a contract to make ten tubes. Raytheon president Marshall then made a bet-the-company decision by investing in a new building and the special equipment required for the process, including a hydrogen oven.[12] Within a month, Raytheon was making thirty magnetrons a day, twice Western Electric's output. With Spencer's promise fulfilled, the contracts started to roll in. Before long, the company was manufacturing the majority of the magnetrons for American and British forces. By the end of the war, Raytheon was pumping out nearly 2,000 magnetrons a day,[13] about 80 percent of all the devices used by the Allies.[14] Spencer and Marshall's gamble had paid off handsomely. In 1945 Raytheon pulled in revenue of $180 million, a staggering jump from $1.5 million before the war.[15]

More important, the gamble paid huge dividends for the Allies. From early 1941, when the new magnetron-powered detection system began to be installed, British and American planes had air superiority over their German rivals. The new system, dubbed "radio detection and ranging" or "radar," persuaded Hitler to permanently cancel his already-delayed invasion. Radar ultimately saved an inestimable number of lives. During the first two years of the war, German bombs killed more than 20,000 London residents. In 1942, after radar had been fully installed, the number of fatalities plummeted to a mere twenty-seven.[16] The scale of the horror experienced in Coventry in the fall of 1940 was never seen again in Britain. The country's remaining architectural treasures, including the massive Gothic edifices of Wells Cathedral and Winchester Cathedral, escaped the war largely unscathed; thanks to the RAF's secret weapon, England's storied past survived to be admired by future generations. A new, modernized Coventry Cathedral, also dedicated to St. Michael, was built right next to the old one after the war, becoming the city's third cathedral.

In the later years of the war, Raytheon expanded beyond magnetron tubes into building whole radar systems, which were then installed on American ships in the Pacific. "With radar we could see the Japanese warships at night," says Raytheon archivist and former vice-president Norman Krim, who has been with the company in various executive roles since its beginning. "They had no idea we could see them and that turned the war around."[17] Vannevar Bush shared that view in his memoirs, where he wrote that radar's importance to ending the war was surpassed only by the atomic bomb.[18] James Phinney Baxter III, the official historian of the U.S. Office of Scientific Research and Development, was no less effusive: "When the members of the Tizard Mission brought the cavity magnetron to America in 1940, they carried the most valuable cargo ever brought to our shores."

The magnetron's military impact is hard to overstate. The scientists who developed radar had an easy moral justification: they were working on a defense system for an unjust war fought against an evil enemy. As with all technology, however, radar also had its dark side. Just as it saved thousands of lives, it also helped end many more. Radar guided the *Enola Gay* to its destination, Hiroshima, where it dropped the atomic bomb that killed an estimated 140,000 people, and helped *Bockscar* find Nagasaki, where another 80,000 were killed by the second bomb.[19] Radar has been installed in every guidance system, fighter jet and bomber used in every war since, bringing its total death count to date to an inestimable figure. Journalists who hailed the invention as "our miracle ally" in 1945 also correctly identified radar's dual nature by tracing it back to its roots. "In a very real sense it represents the mythical death-ray by giving accurate precision so that the death stroke may be delivered," said a *New York Times* editorial.[20]

Radar in the Kitchen

When the war ended, Raytheon's fortunes sank just as fast as they had climbed. The American government had ratcheted up its defense spending as the war progressed, devoting almost 90 percent ($82.9 billion) of its entire 1945 budget to military expenditure. The following year, that spending decreased dramatically to just over three-quarters of the total budget, then plummeted to 37 percent ($12 billion) in 1947.[21] Raytheon was scrambling. At the end of the war the company had employed 18,000 people but was down to 2,500 by 1947. Profit dropped to $1 million by 1956,[22] from $3.4 million in 1945.[23] Krim, a young engineer at Raytheon in its early days, remembers how dismayed Percy Spencer was. "He said, 'What the hell am I going to do?'" Krim recalls. "No more war, no more radar, no more magnetrons. 'I've got to find some use for these magnetrons to keep these people working.' There was a mad rush for products we could make."[24]

Raytheon was able to sell radar devices to commercial shipping operations, including public services such as ferries, but needed to put its invention to commercial use if it was going to stay afloat. The company's first real foray into the wider consumer market was the poorly thought-out Microtherm, a gadget that used the heating properties of the magnetron to treat a variety of ailments, including bursitis and arthritis. The equipment, sold only to doctors, medical suppliers and institutions, could heat "any area, allows temperature penetration of as much as two inches and increases blood circulation by 250%," according to news reports at the time.[25] As smart as Spencer was, frying away aches and pains with microwave radiation was simply not one of his better ideas. Doctors in the forties and fifties agreed, and the Microtherm sold poorly. Krim, who by the sixties had risen through the Raytheon ranks to become the company's "undertaker"—the person called in to dispose of unwanted assets—sold off the money-losing Microtherm business in 1961.[26]

The magnetron's ultimate commercial use was found by accident. Near the end of the war, Spencer was experimenting with magnetrons in his lab and noticed that a chocolate bar in his pocket had melted. Curious about the device's heating effects, he brought in some popcorn kernels, which popped after being exposed. The next day, he exploded an egg using its heat waves. (I remember making the same discovery at age eight, when I blew up an entire pack of hot dogs in our microwave, much to the dismay of my screaming mother.) Spencer knew he was on to something so he applied for and got a patent on microwave cooking. A team of engineers set to work on transforming the magnetron into a cooking device and before long, their efforts bore fruit: they created an oven that heated the water molecules in food but left moisture-free ceramic or plastic containers cool.

The first microwave ovens were hulking behemoths. They stood over 6 feet tall, weighed over 600 pounds, and were the size of a refrigerator. They weren't cheap, either; Raytheon sold them mainly to large restaurants, hotel chains, ocean liners and railways for between $2,000 and $3,000, or the equivalent of $22,000 to $34,000 dollars in 2010 terms.[27] They were made of solid steel with lead-lined ovens to prevent the microwaves from escaping. Their name, however, was perfect: the Radarange.

Large industrial customers loved the Radarange because it dramatically cut down on cooking times. It cooked a potato in four minutes, a ten-ounce sirloin steak in fifty seconds, hot gingerbread in twenty seconds and a lobster in two-and-a-half minutes. The highly competitive steamship industry—where cruise liners emphasized speed, style, luxury and, above all, cuisine—particularly prized them. Potato chip makers such as Lay's also greatly preferred the microwave ovens to their traditional infrared counterparts for drying chips that had just been cooked in oil. Drying with infrared ovens took days while the Radarange did the trick in minutes.

Raytheon tried to expand its market with the first Radarange for the home in 1955, but its enormous expense—about $1,200, or the equivalent of $9,000 today—meant few sales.[28] There were also Microtherm-like safety concerns; many families weren't sure if they wanted to be near a radiation-emitting device. By 1957 only a few thousand had found their way into American homes.[29] Five years later, the ovens had dropped in price to just under $800, but that was still beyond the means of most families, and only 10,000 units had sold.[30] Still, some consumers recognized the irreversible hand of progress when they saw it. "This is not a trend," one housewife said. "The only thing I don't cook in my electronic range is coffee. It is a time saver because I can prepare dinner in a half an hour."[31]

Raytheon's new president Thomas Phillips shared that sentiment, even though the company had lost millions on the Radarange by 1965. He felt the only way to get a return on investment was to speed the oven's adoption in the home, so he acquired Iowa-based Amana Refrigeration and transferred Raytheon's knowledge of microwave ovens to the freezer maker. Krim recalls that Amana president George Foerstner's plan to spur sales was simple. "He said, 'I don't give a damn what's inside that box, it has to sell retail for less than $500.'" The home-appliance maker succeeded where the military contractor failed—by squeezing production efficiencies into the manufacturing process. Amana not only brought the Radarange's price down to under $500, it also shrank the oven to fit on a countertop. Helped by government safety regulations that assured consumers the ovens were safe, sales boomed. Estimates pegged sales of microwaves in 1975 at 840,000, with Amana predicting that 10 percent of American homes would have one by early 1978.

The ovens took off even faster in Japan, where safety concerns were less prevalent; about 1.5 million were sold in 1975, representing about 17 percent of households.[32]

The secret behind the ovens' success was best summed up in a 1976 *New York Times* article. Estelle Silverstone, a New York attorney whose husband was a radiotherapist, was quoted reflecting on the new reality facing women—that of a double-income, dual-career family, short on time for meal preparation. "I've had a microwave for seven years. I don't think I could live without it," she said. "Leftovers don't taste like leftovers anymore. I hate to clean up and there are no pots and pans. It's not a substitute for a conventional oven, but I find it indispensable."[33]

The age of cooking convenience had finally arrived in the home. The microwave oven was the perfect invention for a post-war society that put a premium on speed. With an increasing number of families seeing both parents going off to work, spare time was becoming more and more precious. The microwave helped facilitate those busy lives.

By the early twenty-first century, 96 percent of American homes[34] and 87 percent of British homes had a microwave oven.[35] Today, about 350 million are in use around the world and another twenty million are sold each year. The microwave has reached such a level of ubiquity that it is no longer considered the iconic aspirational purchase it once was. In Britain, where the magnetron was invented, the microwave was removed from the basket of goods used to measure the cost of living in 2008. The plummeting value of the ovens, which can now be had for as little as $25, no longer provides a useful indicator of consumer trends. "We have to make room for new items in the basket and microwaves are no longer different to any other household appliance," a British statistician said.[36] The microwave's invasion of the home is complete, with the previously high-tech device now as mundane as a toaster or can opener.

The Microwave's Sidekicks

The Radarange didn't revolutionize home cooking on its own, though. It had lots of help in the form of new plastics such as Teflon and Saran that were also side effects of weapons development. Teflon, for one, was a direct by-product of the Manhattan Project.

In 1942 U.S. Brigadier-General Leslie Richard Groves, the military commander of the atomic bomb project, twisted the figurative arm of chemical and explosives maker DuPont to help. The company had wanted to steer clear of the conflict after being accused of profiteering during the First World War for selling munitions to Britain and France before the United States joined in. DuPont accepted Groves's task reluctantly and limited itself to an official fee of one dollar[37] after the general argued that the bomb would shorten the war and prevent tens of thousands of American casualties.[38] His argument was likely strengthened by the fact that President Roosevelt's daughter-in-law Ethel was also the DuPont family's heiress. Appearances aside, DuPont took on the key responsibility of producing plutonium, the man-made element derived from the chemical separation of uranium atoms. The company embraced the mission with zeal and selected Hanford, a small, remote mountain town along the Columbia River in Washington State, as the site of its main production facility. By late 1944, after an investment of several millions toward building chemical reactors, separation plants, raw material facilities, acres of housing and miles of roads, the once desolate town had grown to become the third largest city in the state, with a population of 55,000.[39] Hanford was in fact the largest plant DuPont had ever constructed.[40]

Plutonium production was a laborious and expensive process that required miles upon miles of pipes, pumps and barriers. An ounce of dust, grime or grease could ruin the entire system by entering through a tiny pinhole, yet a sealant that could perform a perfect patching job did not exist. DuPont decided to try out a substance that research chemist Roy Plunkett had accidentally discovered in 1938 at one of its labs in New Jersey. While experimenting with refrigerants, Plunkett opened a container of tetrafluoroethylene, only to find that the gas inside had solidified into a white resin. He found the new substance, which he dubbed polytetrafluoroethylene, to be extremely slippery and resistant to chemicals and heat. DuPont tested the substance as a sealant in its plutonium plant and found it plugged all the pipes and pumps perfectly. It was also put to use as a non-corrosive coating for artillery shell nose cones and as a high-performance lining for liquid fuel storage tanks, tasks at which it also excelled. The company patented the substance in 1941 and trademarked it just before the war ended under the name Teflon.

The substance was first sold in 1946 as a sealant for the electronic, chemical and automotive industries and took off in the late fifties once a home use was found. In 1954 French engineer Marc Grégoire invented a process for bonding Teflon with an aluminum frying pan, with which he launched his Tefal company. Consumers, happy about no longer having to fry their food in a pound of butter to stop it sticking to the pan, snapped up Tefal's product (and the inevitable clones) in droves. By 1961 the company was selling a million pans a month.[41] Teflon's use expanded again in 1969, when American engineer Bob Gore discovered it could be stretched into a porous, super-strong filament. His new version of Teflon turned out to be an excellent transmitter of computer data and a good material for surgical supplies. Its ability to keep out moisture but let in air also meant it was the first material that could actually "breathe," which made it ideal for waterproofing clothing. After several years of development, "Gore-Tex" clothes hit the market in 1980, and skiers would never again have to come home soaking wet.

Saran Wrap also had its origins during the war, and like so many good inventions, it too was an accident. In what reads like the origin story of a comic-book super hero, Ralph Wiley, an unwitting college student working at Dow Chemical's labs in Michigan, was performing his chores one night when he found some beakers he couldn't scrub clean. He dubbed the green substance stuck to them "eonite" after an indestructible metal that was supposed to save the world from the Great Depression in a *Little Orphan Annie* comic strip. Upon examining the goo, Dow researchers gave it the more scientific name of polyvinylidene chloride (PVDC). Wiley didn't end up gaining super powers, but Dow did turn the substance into a greasy green film, which was dubbed Saran, and tested it during the war by spraying it on fighter planes. Saran did a good job at keeping out oxygen and water vapor and was perfect for protecting the planes on aircraft carriers from the spray of salt water. The substance saved the navy time and effort by allowing planes to be shipped on the decks of aircraft carriers, rather than disassembled and stored below decks in pieces. Guns were also wrapped in the protective plastic, like death-dealing lollipops. A wartime ad from Dow proclaimed that when "men on our fighting fronts throughout the world . . . unpack a machine gun they find it protected from moisture with Saran Film. There are no coatings of grease to be removed—no time lost. The gun slips out of its Saran Film envelope clean, uncorroded, ready for action!"

After the war, Saran went commercial. By 1950 the plastic was being sprayed onto everything from bus seats to clothes to drapes, all of which it made more water-resistant. Dow's revenue climbed steadily thanks to Saran[42], and by 1952 the company was churning out more than fifty million tons of the stuff.[43] The plastic's real impact, however, came a year later, when it was turned into a clear, clingy film that could be stretched over food, allowing people to store leftovers in the refrigerator. Saran Wrap was the perfect partner for the Radarange; the plastic kept leftovers from spoiling long enough to be reheated in the microwave. Buoyed by sales of new plastic products, particularly Saran Wrap, Dow ended the decade with record sales and profits.[44] Saran and other competing plastic wraps were becoming ubiquitous kitchen fixtures.

The most important plastic to come out of the war, however, was polyethylene. This highly versatile, variable-density substance was discovered, again by accident, before the war by British researchers working for Imperial Chemical Industries in London. In their search for new plastics, Eric Fawcett and Reginald Gibson found that a mixture of ethylene and benzaldehyde produced a white, waxy substance. The experiment yielded a usable result only because it had been contaminated with an unknown amount of oxygen, an accident that took the scientists years to recreate. Polyethylene wasn't truly born until 1939, just in time for the war, when it became the primary material for insulating cables on radar. Although Germany developed its own detection system during the war, its scientists never did come up with polyethylene, which meant its troops faced a disadvantage in situations where moisture was a factor. German boats and planes traveling through rain and clouds often saw their radar malfunction, proving the historical significance of a seemingly mundane type of plastic.

DuPont licensed polyethylene from ICI, but aside from insulating radar and other telecommunications equipment, the company didn't know what else to do with it.[45] Late in the war the copmany gave one of its former engineers, Earl Tupper, a few tons of the plastic to play with, which he used to make gas masks and signal lamp parts. Tupper struck it rich after the war by using the plastic to create a range of storage containers with liquid- and airtight sealable lids.[46] The containers, egotistically dubbed "Tupperware"—not that Gore-Tex was particularly modest—showed DuPont and others the way in plastics. Before long, polyethylene was everywhere: in dishware, furniture, bags, toys (including two of the biggest crazes of the fifties, the Frisbee and the hula hoop), shampoo and soda pop bottles, packaging, pens and even clothes. The plastic's versatility and uses were limited only by manufacturers' imaginations. ICI and DuPont proved to be the most imaginative and established themselves as the biggest plastics makers in the world. For his part, Tupper sold his Tupperware company to Rexall Drugs in 1958 for a reported $10 million, which he used to buy a small Central American island where he lived as a hermit until his death in 1984. This only sounds odd until you compare it with the fate of Richard James, the inventor of the Slinky, as we'll see in chapter four.

The Dark Side of Plastic

For all the Allied advances in plastics, it was actually Germany that led the way in the development of synthetic materials. The Nazis had two good reasons for their accelerated research: Germany had experienced material shortages more acutely than any other nation during the First World War, and after that conflict it was prohibited by Allied sanctions from stockpiling resources that could be used for armaments. As a result, the German people were already familiar with synthetic or "ersatz" products by the twenties. In 1925 a group of chemical companies were brought together into the Interessen-Gemeinschaft Farbenindustrie conglomerate, better known as I.G. Farben, as part of a strategy to create materials that could circumvent treaty limitations and allow Germany to prepare for future wars. Over the next decade, the conglomerate hired the country's best scientists, who then rewrote the book on polymers—chemical compounds made up of a large number of identical components linked together like a chain. With their government-approved mandate, Farben chemists synthesized an average of one new polymer every day for ten years.[47] When the Nazis came to power in 1933, Hitler immediately recognized the value of plastic and put Germany's scientific community at the disposal of the state. "It is the task of science to give us those materials nature has cruelly denied us," he said. "Science is only free if she can sovereignly master those problems posed to her by life."[48] By the time the war broke out in 1939, Germany's military machine was largely synthetic, and more than 85 percent of its strategic materials were being made by Farben.

Hitler knew Germany would be cut off from two key resources—oil and rubber—once the war began, so he urged Farben to come up with synthetic alternatives. The results were two big hits: a hydrocarbon made by mixing carbon dioxide, carbon monoxide and methane, which German tanks and other vehicles could use as fuel; and a new, plastic type of rubber. The synthetic rubber was created by bonding together two polymer compounds, butadiene and styrene, into a so-called "co-polymer." Farben called its new substance an elastomer for its elastic properties and officially dubbed it Buna, a contraction of butadiene and Na, the chemical symbol for sodium, which was the catalyst for the polymer reaction.[49]

Farben's most notorious invention, however, was Zyklon B, the pesticide used by Nazis to gas concentration camp victims. The chemical conglomerate's brain trust profited heavily from its association with the Nazis and their concentration camps, both through the slave labor the camps provided and, even more horrifically, from the bountiful test subjects. At its peak, Farben's factory at Auschwitz in Poland alone made use of 83,000 slave laborers and an undocumented number of unwilling human guinea pigs.[50] The conglomerate and many of its most zealous scientists faced justice in the Nuremburg trials that followed the war (thirteen of its directors were found guilty of war crimes and served prison time), but much of the work lived on through component companies after Farben was dismantled in 1951. Several of today's largest multinational firms owe part of their post-war successes to the often ill-gotten intellectual property inherited from Farben, including film manufacturer Agfa-Gevaert, chemical maker BASF and pharmaceutical companies Sanofi-Aventis (derived from a merger of Farben spinoff Hoechst and France's Rhône-Poulenc Rorer) and Bayer.

Plastics also caused their share of physiological damage. Through the fifties and sixties, research showed that workers producing synthetic substances were vulnerable to developing a host of medical conditions, including heart arrhythmia, hepatitis, gastritis, skin lesions, dermatitis and cancer. Worse still, some plastics—particularly polyvinyl chloride (PVC) and polystyrene—were found to be able to leach into food and cause cancer in unsuspecting consumers. By the seventies, the public was wary of plastic and companies using it began to feel the backlash. Coca-Cola, for one, was set to introduce the world's first plastic soft drink bottle in 1977, but had to pull the plug amid fears that the product could cause cancer. Coke's bottle, made of acrylonitrile styrene, was designed in conjunction with chemical giant Monsanto at an estimated cost of $100 million. Monsanto's product application stated that low levels of about fifty parts per billion of carcinogenic plastic "may form and migrate into the beverage"—a negligible amount, but it was enough for the Food and Drug Administration to reject the bottle.[51] Monsanto closed its bottle manufacturing plants and Pepsi beat Coke to the punch with a polyethylene bottle, designed by DuPont, that passed FDA muster.

Coke's first attempt at a plastic bottle—it eventually introduced an FDA-approved product in 1978—was just one casualty of the public's growing unease with plastics, a discomfort that became part of the battle between consumer advocates and food processors that continues today. In the past few years, for example, health authorities in several countries have banned the plastic Bisphenol A from baby bottles because tests have linked it to cancer and hormone imbalance.

In addition to health risks, plastics are also strongly associated with environmental damage. Most plastics degrade very slowly, which means that the ketchup bottle in your fridge is likely to be around long after Armageddon. By the eighties, this presented a huge problem for overflowing garbage landfills around the world. Late in the decade, large corporations started to feel pressure from consumer groups to limit their plastic waste output and institute recycling programs. In 1987 one group, the Citizens Clearinghouse on Hazardous Waste, found that McDonald's alone was contributing more than a billion cubic feet of foam packaging waste each year.[52] Along with another grassroots group, the Vermonters Organized for Cleanup, the CCHW pressured the fast-food chain into switching to recyclable paper packaging in 1990, a move McDonald's said would reduce its waste output by about 90 percent.[53] For environmentalists, the chain's switch was a major win, but it was only a small victory in the battle against an overflowing tide of permanent, non-biodegradable waste, a struggle that continues today.

But health and environmental concerns were the furthest thing from the collective minds of people in the fifties and early sixties. They had made it through the worst economic crisis and wars in human history and they wanted a chance to stretch their legs and live it up. As the *Life* ad said, after total war came total living. The microwave oven freed up people's time from chores like cooking, while plastics provided a veritable cornucopia of new products on which hard-earned salaries could be spent. These consumer goods kicked off a new lifestyle, one dedicated to instant gratification, prosperity and indulgence, the diametric opposite of life during the thirties and forties. These weapons of mass consumption, derived from inventions that helped perfect the art of killing, forever changed daily life and paved the way for the sort of total living that would come in later decades.

BETTER EATING
THROUGH CHEMISTRY

McDonald's custom designed a condiment dispenser to squirt a precise amount of ketchup and mustard. This dispenser is now on display at the unofficial McDonald's museum in San Bernardino, California.
PHOTO COURTESY OF THE AUTHOR.

Do you know what breakfast cereal is made of? It's made of all those little curly wooden shavings you find in pencil sharpeners!

—ROALD DAHL, *CHARLIE AND THE CHOCOLATE FACTORY*

Like McDonald's restaurants everywhere, the chain's outlets in Hawaii sell a truckload of Egg McMuffins and hash browns every morning. But they also go through a large amount of something few other McDonald's even sell: Spam. Aside from the usual breakfast fare, Hawaiians can choose from several items that incorporate the canned meat, the most popular being the Spam, eggs and rice platter. The meal features a fried egg, like the one found on a McMuffin, with rice and a cooked slice of the meat. The chain's biggest rival, Burger King, also offers a Spam platter of rice and scrambled eggs with two slices of the meat. For both companies, having Spam on the menu is a no-brainer, because Hawaiians love it. Despite having a population of only 1.2 million people, Hawaii leads the United States in Spam consumption, with more than seven million cans a year, or about six per person.[1] McDonald's sells more than 3,000 Spam items in Hawaii every day and has expanded the offering to other Pacific islands, including the Philippines, Saipan and Guam.[2] Guamanians even put the Hawaiians' heavy consumption to shame by eating 2.5 million cans a year, or sixteen per person.[3]

Why do Pacific Islanders love a food that is reviled by so many others? A food so hated that when it came time to name annoying unwanted email messages, nothing but "Spam" would do?

The love affair started during the Second World War, when the American military imported tons of the canned meat to the islands. To properly understand Spam, though, we have to go back several centuries. Despite being commonly derided as a poor-quality meat, Spam is actually a brilliant application of food technology that traces its lineage back to Napoleon. The French emperor, who famously said that an army marches on its stomach, was faced with the problem of keeping his soldiers fed and well nourished while in the field. In the eighteenth century, a time before any significant food preservation and processing, this was a big challenge. So in 1795 Napoleon announced a contest: he would personally award 12,000 francs, a king's ransom at the time, to anyone who could invent a food-preservation technique that would help feed his troops. Many tried and failed until Nicolas Appert, a confectioner from a small hamlet east of Paris, answered the call.

In 1810, Appert discovered a process whereby food, generally meat or vegetables, could be preserved when it was sealed in a glass jar, covered with canvas and then boiled. The process greatly lengthened the shelf life of the food by sealing it in its own juices, which emerged during cooking. Appert was not a scientist and didn't understand the principles behind his discovery, but he knew it worked. His find won him the reward and the emperor's praise and was soon refined by fellow Frenchman Pierre Durand, who used a tin can instead of a glass jar, which cooked and preserved just as well but had the added bonus of extra durability for transport. Even so, Appert became known as the father of canning and the inventor of modern food preservation and processing. From then on, the history of food processing evolved in virtual lockstep with the history of war.

Appert's heir apparent emerged more than a century later across the Atlantic. In 1910 Minnesota-based Hormel was a mid-sized meat-packer with distribution centers in five states and an exporting business to Britain. Jay Hormel, son of founder George, earned his stripes during the First World War, when he served in France as a quartermaster responsible for supplying troops with clothes and provisions. When his superiors complained about the time and effort it took to ship meat across the Atlantic, he came up with a simple suggestion: rather than pack entire sides of beef, why not debone them first? Hormel flew back to the packing plants in Chicago to demonstrate and, before long, small packages of boneless beef were being shipped to Europe, saving time and money. Most important, the troops were happy because they got to eat meat more frequently.

After the war, Hormel went to work developing new products for his father, including the first canned ham, using techniques similar to Appert's. By 1929 he was company president. His big hit, however, came in 1936 while he was searching for a use for pork shoulder, a pig part that was not selling well. Pork shoulder meat, when removed from the bone, came in small chunks that consumers didn't like—they were used to large chops, the bigger the better. Removing the meat from the bone was also expensive and time-consuming, which compounded the problem. Hormel experimented with additives to make the shoulder pork tastier. Like a modern-day Doctor Frankenstein, he combined it with different parts of the pig before settling on ham, which comes from the animal's rear thigh. When the two meats were ground together and mixed with water and salt, they formed a pink paste. Hormel squirted the concoction into a twelve-ounce can, sealed it and cooked it using the Appert method. The result was a tangy ham-like meat. After adding sodium nitrite, a powder added to many meats to prevent them from turning an unappetizing grey, Hormel had his masterpiece. He called it Hormel Spiced Ham, a versatile canned meat that could sit on the shelf for years and not go bad. (The only thing missing from this Frankenstein-like scene was the mob of pitchfork-wielding villagers outside the door.)

His creation, however, didn't sell. People were suspicious of meat that came in a can. In Napoleon-like fashion, Hormel held a contest at one of his parties to come up with a new name. Kenneth Daigneau, a New York actor and brother to one of his company's vice-presidents, won bragging rights and a crisp $100 bill for thinking to contract "spiced ham" to Spam. Under the more marketable name, the company ran a flood of magazine and newspaper ads that touted the meat's versatility. "You've never known a meat like Spam, Hormel's miracle meat of many uses for many occasions," one read. "Slice it cold for Spamwiches, salads, snacks. Serve it hot as Spam and eggs or baked Spam. Any way you eat it, Spam hits the spot."[4] Thanks to the marketing blitz, the product enjoyed modest sales until the war broke out. Then it really took off, which was ironic given that Hormel was fiercely opposed to the United States joining the conflict.

Naturally, Hormel soon changed his tune. As one of the larger meat packers in the United States, with military connections and export ties to the United Kingdom, he had an easy time landing a job as a major supplier to the United States–Britain Lend-Lease program. Aside from weapons and ammunition, the United

States had provided Britain with more than a billion dollars in aid by 1941, including food.[5] Spam was selected by the U.S. military as the perfect war food—portable, light, cheap and virtually unspoilable—and it was shipped to Britain for consumption by soldiers and civilians alike. After the United States entered the war, tons more were shipped to troops stationed in Hawaii and other Pacific islands, where it was packaged into rations. By 1944 about 90 percent of Hormel's Spam output was going to military forces, doubling the company's overall revenue. Hormel was gobbling up so much tin for cans that supplies for home use were curtailed and civilian Americans were forced to use glass jars.[6] By the end of the war, civilians and soldiers worldwide had consumed 100 million pounds of Spam, or 113 million cans.[7]

Soldiers had a love-hate relationship with the meat. Some, who were forced to eat it for breakfast, lunch and dinner, despised it as "ham that failed its physical." Others were more appreciative. "I know there were a lot of jokes about Spam," recalled one infantryman, but "it probably saved lives in the field. It was easy to transport and it could last for a long time."[8] Despite the revulsion, Spam remained popular after the war. Sales reached one billion cans by 1959 and by 2007 it was available in more than a hundred countries.[9] Pacific Islanders, meanwhile, adopted it as their native cuisine. For many, Spam is the ideal "comfort food." Indeed, Spam *musubi*—a strip of the cooked meat sitting on a block of rice, held in place by a sushi-style band of seaweed—was sold in Hawaiian 7-Eleven stores long before McDonald's thought to get in on the action.

By using simple chemistry in combination with a proven Napoleonic-era preservation technique, Jay Hormel had come up with the original "Frankenfood." He showed his peers that easily spoiled natural foods could be made more durable with the application of some basic science. Moreover, just as Raytheon had shown other electronics makers how to get rich from government contracts, Hormel too had cashed in by supplying the military at home and abroad. Spam showed other foodmakers what such contracts could mean to their bottom lines.

The Long, Long March to Nutrition

Spam is not and never has been a healthy food. A small, two-ounce serving contains about one-third of the daily recommended total fat and sodium intake. Too much Spam can raise your blood pressure and make you fat. Back in the forties, however, what little information the U.S. military had on Spam's nutritional quality was set aside in the quest for quick, inexpensive and durable meat. During the war, the Food and Nutrition Board—formed in 1941 to investigate issues that affected national defence—developed the Recommended Daily Allowances (RDA) list to spell out the components of a healthy diet. The list, which was influenced by wartime shortages of certain foodstuffs including meat, was first published in 1943 and revised numerous times afterward. Regulations requiring processors to disclose what percentage of the RDA their foods provided, however, were a long time coming because the industry fought such moves every step of the way.

In the early twentieth century there were few rules governing food; American processors were subject only to the Pure Food and Drug Act of 1906, administered by the Department of Agriculture's Bureau of Chemistry. The act prohibited the manufacture and sale of poisonous and "adulterated" foods, a vague definition that applied to fillers of reduced quality or strength, coloring to conceal damage or inferiority, injurious additives and the use of "filthy, decomposed or putrid" substances.

In 1927 the bureau was reorganized into the Food, Drug and Insecticide Administration, which dropped insecticides from its name in 1930 to finally become the FDA. The 1938 Federal Food, Drug and Cosmetic Act did little new other than expand the 1906 legislation to include cosmetics and authorize factory inspections and spell out acceptable food colorings. It wasn't until 1958 that significant food safety requirements were established, when the Food Additives Amendment banned the use of additives found to cause cancer in humans or animals. A similar amendment two years later outlawed cancer-causing colorings. The Fair Packaging and Labeling Act of 1966 required that all consumer products be honestly and informatively labeled, while 1975 produced a landmark FDA ruling that finally required processors to display nutrient information on their foods. The new rule, however, only applied to products that made nutritional claims and was not expanded to include all foods for nearly two decades, in 1992. Regulations in other developed countries followed similar trajectories and timelines.

By that time, however, Pacific Islanders were thoroughly hooked on Spam. Like Native American populations afflicted by the diseases brought by colonizing Europeans, the Islanders were exposed to the unhealthy food imported by American armed forces. Spam and other canned meats have had a devastating effect on the region in what has been called a "raging epidemic" of diabetes, stroke and heart disease. Eight of the world's ten most obese countries are Pacific islands. Nauru, northeast of the

Solomon Islands, fares the worst with 95 percent of inhabitants over the age of fifteen considered obese by the World Health Organization. "What is unfolding here is a physical disaster and a fiscal disaster," says Carl Hacker, the director of economic policy and planning in the Marshall Islands.[10] Experts say Spam and other fatty canned meats such as corned beef are squarely to blame. But for the rest of the world, Spam has proved to be only a small contributing factor to obesity, when compared with some of the other food products that came about as the result of war technology.

Potato Diplomacy

By the time the Second World War broke out, John Richard Simplot had put Idaho on the map as the potato heart of the United States. His company was already the largest shipper of fresh potatoes in the country and ran a side business selling dried onions, which he dehydrated in a modified prune dryer. Like sides of beef, potatoes were difficult to ship to troops because of their weight and tendency to spoil. But Simplot, who had dropped out of school in the eighth grade and gone into business for himself at fourteen, was every bit as inventive as Jay Hormel and knew the solution lay in processing. He took his cue from the ancient Incas, who were the first to grow potatoes, in the high Andes more than 4,000 years ago.[11] In between herding llamas and getting high chewing coca leaves, the Incas discovered that storing potatoes at high altitudes caused them to lose much of their moisture, which increased their longevity. Simplot's scientists discovered that the higher air pressures and lower temperatures of the Andes enabled a natural form of freeze-drying, an effect they then duplicated using their prune dryer back in Idaho. They took the resultant dried spuds and diced them into flakes, which could then be reconstituted into mashed potatoes by adding water or milk. Simplot packaged the flakes in boxes and sold them to the army, a move that ended up making the already-rich Idahoan wealthy beyond his wildest dreams. After the war, he bought his own potato farms, cattle ranches and fertilizer plants, as well as lumber mills and mining claims.

As with radar, demand from the army for dehydrated potatoes dried up after the war, leaving Simplot to find new lines of business. Clarence Birdseye, a Brooklyn native, had pioneered flash-freezing in the twenties but his fish products sold poorly because few grocery stores—and fewer homes—owned freezers to store them. That changed during the war when shortages of fresh foods led to a boom in refrigerator and freezer sales.

The first successful frozen food was orange juice, invented by the Massachusetts-based National Research Corporation during the latter part of the war. NRC's early attempts to dehydrate orange juice by boiling the water out of it didn't work because the process destroyed the flavor. The company's founder, Richard Morse, found a suitable alternative with a high-vacuum process he had created to dehydrate penicillin, blood plasma and antibiotics. Morse was able to suck the moisture out of the juice using a giant vacuum machine to create an orange powder which, when reconstituted with water, retained some of its flavor and vitamins. The breakthrough won him a hefty contract with the army in early 1945.[12] NRC formed the Florida Foods Corporation and began building a plant in Florida in the spring, only to have the army cancel its contract when the war ended that summer. The company scrambled to reorient itself to the consumer market, renaming itself Vacuum Foods and beginning to sell frozen orange juice concentrate, an intermediate step in its process. The slushy concentrate was more expensive to produce than the powder, but it also produced a more real-tasting juice. A Boston marketing firm came up with the name "Minute Maid," a brand that reflected the amount of time the juice took to make and which referenced the heroic Minute Men militia of the American revolutionary war. The company's frozen concentrate hit stores in 1946 and sales were immediately good, then exploded over the next few years. Vacuum Foods saw revenue go from under $400,000 in its first year to nearly $30 million just five years later.[13]

The company changed its name once again, finally settling on Minute Maid in 1949. Frozen orange juice proved to be immensely successful, and by the end of the decade Minute Maid was competing for space in grocery stores against sixty different imitation labels. Company president John Fox proclaimed that Minute Maid had single-handedly saved the Florida citrus industry—in 1946 it was appealing to the government for subsidies, but by 1950 it was overwhelmed with demand for oranges.[14] Powdered orange juice, meanwhile, resurfaced in several forms during the fifties, including Tang (1959), but many of these products were full of sugar and other additives to make up for the taste lost in processing.

Minute Maid's success was not lost on Simplot, who had scientists trying to figure out a way to freeze french fries. After a number of attempts that resulted in poor-tasting and mushy fries, one of his researchers, Ray Dunlap, discovered a method that kept the potatoes' flavor intact. Dunlap first pre-cooked the fries in hot oil for two minutes, then immediately blasted them with super-cooled air. This flash-freezing technique brought the temperature of the fries down to around −30° Fahrenheit in a manner of minutes and had a major advantage over previous attempts. Slow freezing over a longer period of time caused the water molecules inside the fries to expand, which made the fries mushy when they were eventually thawed. Flash-freezing, on the other hand, didn't give the water the opportunity to gather, so when the fries were thawed they retained the same moisture as before. The result was a frozen fry that, when re-cooked in oil for another two minutes, tasted the same as a fresh one. Dunlap presented his invention to Simplot, who took one bite and said, "That's a helluva thing."[15]

Simplot sold the frozen fries to grocery stores in 1953, but they didn't take off right away. Although the fries could be cooked in any home oven, they tasted best when made in hot oil, a method few households were equipped for. The trick lay in the moisture inside the fries—hot oil caused the water molecules to evaporate, leaving gaps that were then filled by the oil itself, which added to their taste. Simplot sought out restaurants that were equipped with oil cookers, such as the fast-rising McDonald's. He was already a major supplier to the chain, accounting for about 20 percent of its potatoes. He took his idea to McDonald's president Harry Sonneborn, who gave him a frosty reception. "He laughed at us," Simplot said. "The only thing he was interested in talking about was fresh potatoes."[16]

Still, potatoes posed a big problem for McDonald's. Because of their high solid content, the Idaho Russets the chain used were available fresh only nine months out of the year. They were harvested in the fall and kept in cold storage throughout the winter, but tended to go bad during the hot months, so the restaurant chain was forced to switch to California white potatoes during the summer. The white potatoes didn't produce as crisp a fry, though, which gave McDonald's founder Ray Kroc quality-control headaches. Fresh fries were also McDonald's most time-consuming menu item to prepare since the potatoes needed to be peeled, cut and cooked. Finally, the fast growth of Kroc's chain—it had about 725 restaurants at the time and would have more than 3,000 by the end of the decade—was making it difficult to maintain potato uniformity. "The sugar content of the potatoes was constantly going up and down, and they would get fries with every color of the rainbow," Simplot said.[17]

The potato baron used his clout with the chain to go over Sonneborn's head to meet directly with Kroc. Appealing to his obsession with quality, Simplot convinced Kroc to take a chance on frozen fries and, with a handshake, the two changed potatoes forever. Kroc agreed to try the frozen fries and, when customers couldn't tell the difference, McDonald's began a large-scale conversion, a move it completed in 1972. Other fast-food chains followed McDonald's lead, leading to an explosion in consumption. In 1960 the average American ate eighty-one pounds of fresh potatoes and about four pounds of frozen fries. By 2000 that had changed to forty-nine pounds of fresh potatoes and more than thirty pounds of frozen fries, 90 percent of which were bought at fast-food restaurants.[18]

Simplot became immensely wealthy and spread his money into other investments. Some, such as his mining endeavors, did poorly while others, including a $1 million stake in Idaho-based microchip start-up Micron Technology in 1980, paid off handsomely. Micron is now a Fortune 500 company and, in an apparent tribute to the war ties that made Simplot rich, goes out of its way to employ military veterans, whom it actively recruits. In 2006 the company employed more than 3,500 veterans, comprising about 16 percent of its workforce.[19] Simplot also expanded into other frozen foods such as meats and vegetables and even bought Birdseye's companies in several countries, including Australia and New Zealand. His continued expansion and growth was paced by the frozen food industry itself. In the United States alone, frozen food sales had reached $40 billion—one-third of total food sales—by the turn of the new century.[20] In 2007 the industry posted global sales of more than $100 billion.[21]

At the time of his death in May 2008, Simplot and his family were worth more than $3 billion. His legacy to the world will always be the technologically engineered french fry, now a major contributor to what the World Health Organization calls an obesity epidemic. One of the main causes is the "increased consumption of more energy-dense, nutrient-poor foods with high levels of sugar and saturated fats," such as french fries.[22] The rising obesity numbers are not surprising given that one large portion of fries—and in the United States, one in every four vegetables consumed is a fry—constitutes nearly half your recommended daily fat intake.[23]

Powder More Valuable Than Cocaine

Other technological developments during the Second World War helped diminish the overall nutritional value of food. The process of spray-drying, for example, is a method that strips vitamins and minerals from many foods, including milk and eggs. Food processors had been using several different methods to create dried milk since the late nineteenth century, but few met with much success.

Before the Second World War, the most popular method involved a process similar to NRC's juice dehydration technique, where moisture was sucked out of milk, but the costs of running the heavy machinery proved too high, especially given that sales were low. Indeed, dried milk, which came in powder form, was a new substance that was mistrusted, particularly by bakers. When processors first tried to sell milk powder, bakers "looked first at the product, then at the man attempting to make the sale with suspicion. Believing it was some foreign substance to be substituted in place of milk, they would not accept it and thus it required very hard work and education to get them to use it in even a small quantity."[24] As a consequence, by 1939 about one-third of the 600 million pounds of dried milk produced in the United States went to feeding animals.[25] The war changed all that. With a sudden need for milk that would not go bad, demand for the powder skyrocketed and food processors followed suit by investing in the relatively new and expensive spray-drying technology.

The spray-drying process, largely unchanged since its inception, takes several steps. First, the bacteria in the milk are removed through pasteurization, a heating process similar to the method in which Spam is baked. The fat is then removed by spinning or "skimming" the milk in a centrifuge. The pasteurized skim milk is then put into an evaporator silo, where more moisture is removed through further heating. From there, the milk—now about 50 percent solid—finally hits the spray dryer, a large metal cylinder, where it is again heated and blasted with highly pressurized air. The air evaporates whatever water remains and mixes with the milk to form a powder, which then drops to the bottom of the dryer to await cooling and packaging.

The process was relatively new during the Second World War, so the resultant powder still had a chalky taste when rehydrated with water, but it was cheaper, longer-lasting and more efficient than other methods, and it was deemed good enough for bread making. American production of spray-dried milk took off and reached about 700 million pounds by 1945, more than double pre-war levels.[26] Food processors, led by Carnation, found a veritable bonanza in the technology after the war with a slew of new milk-related beverages. Swiss giant Nestlé, which ended up buying Carnation in 1985, released the ever-popular Quik chocolate milk powder complete with spokesrabbit—because everyone knows that rabbits like milk—in 1948, followed by a strawberry version in 1959. By 1954 sales of non-fat dry milk solids had grown from around two million shortly after the war to 120 million pounds, leading some industry observers to wonder if the long-lasting and unspoilable powdered milk might entirely replace its liquid counterpart.[27] (Obviously, it didn't.) Spray-drying wasn't just limited to milk. In 1942 General Foods used the process to create instant coffee, which it supplied to American forces. When the war ended, the company sold the product to the general public as Maxwell House Instant Coffee, which also proved to be a hit.

By the time war broke out, Detroit-based C.E. Rogers was a mid-sized player in the spray-drying industry. Eager to give his company a leg-up over competing milk processors, CEO Elmer Donald Rogers perfected the process of spray-drying eggs. The method was similar but the spray dryer was horizontal and box-shaped rather than vertical and cylindrical, like its milk counterpart, and the eggs weren't heated, because doing so resulted in prematurely scrambled eggs.[28]

Howard Rogers, the company's current president and grandson of Elmer Donald, boasts about how lucrative selling the machines that made powdered eggs was. His company, run at the time by his grandfather, was assigned a priority "second only to the Manhattan Project" for obtaining the materials needed to build spray-dryers, including stainless and carbon steels. "They made a hell of a lot of money around World War Two," he told me. "I know this because my grandfather built a home in Northfield, Michigan, that was monstrous." The move may have paid off handsomely, but the company reverted back to milk production after the war, likely because of the universal revulsion soldiers had for powdered eggs. "Ugh," one naval officer said, "our engineering officer . . . could eat a half dozen of those powdered items and squirt half a bottle of ketchup on them. No one else at the wardroom table could stand to watch him."[29] Larger food processors, including Carnation and Nestlé, furthered research after the war and improved the taste by adding artificial flavors, thereby establishing a commercial market. Today, one in ten eggs is consumed in powdered, frozen or liquid form.[30]

Mass Processing

The advances in dehydration, freezing and drying, however, paled in significance to the advent of the mass spectrometer, a scientific instrument that virtually no one outside of a laboratory has heard of. The spectrometer, which measures the mass and relative concentrations of atoms and molecules, was arguably the most significant technological invention of the twentieth century, next to the atomic bomb. In fact, the bomb itself might not have been possible without the spectrometer.

The field of mass spectrometry was created by Manchester-born physicist Joseph John Thomson in the early part of the century. In 1913 Thomson, who had won the Nobel Prize in Physics in 1906 for his experiments on the conductivity of gases, was investigating the effects of magnetism and electricity on atoms. He believed that if he shot a stream of electron-charged neon gas through a magnetic field, the particles would deviate from their straight path and curve. He used a photographic plate to measure the angle of deflection and his hypothesis proved to be more accurate than he anticipated: the plate showed two different patches of light, indicating that his stream hadn't just curved, it had split off into two different rays. Thomson took this to mean that neon gas was actually composed of atoms of two different weights. He had discovered the isotope: a chemical twin of a natural element that had a different atomic weight than its brethren.

His system was improved upon and by the war had evolved into the mass spectrometer, a device that allowed scientists to accurately identify different molecules and isotopes by their atomic weight. During the war, the spectrometer was used to find uranium isotope 235, the key element in the atomic bomb. The uranium isotope was particularly dense, which gave it maximum explosive power when split. Mass spectrometry, one of two methods of producing elements that could be split, was used at the giant government facility built in Oak Ridge, Tennessee, to create the uranium-based atomic bomb that was dropped on Hiroshima on August 6, 1945. (The second method, used at the DuPont plant in Washington, produced plutonium—a derivative of uranium—through a chemical-splitting process.

The plutonium-based bomb was dropped on Nagasaki three days after Hiroshima.)

After the war, the mass spectrometer was widely adopted by scientists across a broad range of fields and industries, including pharmaceuticals, energy and electronics. Food processors were particularly enthusiastic about the new technology because it took a lot of the guesswork out of their jobs by allowing them to study their products at the molecular level. Scientists could now see how adding one molecule or altering the chemical composition of another affected a given food. In wartime experiments with orange juice, potatoes, milk and eggs, scientists found that processing often deprived food of its taste. The mass spectrometer now allowed them to correct those problems on a chemical level.

This introduced a number of new phenomena to the food industry. First, it led to the birth of the flavor industry. With the capability to mix and match molecules, chemical makers were now able to synthesize any aroma or taste. This was a godsend for food processors, because it meant that they could do whatever they wanted to food and not worry about how the end product tasted—artificial flavoring would take care of that. Not surprisingly, the flavor companies that sprung up saw no shortage of demand from food companies. Today, the global flavor market is worth more than $20 billion and is led by companies such as Swiss duo Givaudan and Firmenich, New York–based International Flavors & Fragrances and Germany's Symrise.[31]

Mass spectrometers also allowed food processors to dissect their competitors' products. If company A came out with a particular food that was a big hit, company B could easily replicate it. Some long-held and zealously guarded formulas, like Coca-Cola's, were no longer secret. Scott Smith, the chair of the food science program at Kansas State University and an expert in mass spectrometers, explains that the device took much of the guesswork out of food production. "If you're looking at coffee and you want to know something about coffee—like are you seeing some differences in the coffee beans from different parts of the world or different types of roasting—you can use taste panels, but you can also use a mass spectrometer to give you more of a subjective analytical approach."[32]

The devices have also become indispensable in ensuring food quality, particularly when something goes wrong. Smith recently used a spectrometer to analyze a chocolate-covered nut product that had been brought to his lab because it "tasted like cardboard." He found heavy oxidization in one of the product's compounds, a problem that was solved by simply eliminating that compound. With food problems, the spectrometer "will usually give you an idea of where to start looking," Smith says. "At my lab, we live and die by it."

The fifties and sixties thus saw an explosion of new food products, many of which were full of new preservatives, additives, flavors and colorings. Pop Tarts, processed cheese slices, Frosted Flakes, TV dinners, Cheez Whiz, Rice-A-Roni, Fruit Loops, Cool Whip, Spaghetti-O's, prepared cake mixes and many other products hit grocery stores and became popular with consumers. Chemical intake shot up dramatically—between 1949 and 1959, food processors came up with more than 400 new additives.[33] The FDA couldn't keep pace; in 1958 the regulator published a list of 200 "Substances Generally Recognized as Safe," but by then more than 700 were being used in foods.[34] In the rush to provide the world with products that wouldn't spoil but tasted just as good through additives, little attention was being paid to the nutritional value or the potential long-term effects of these technologically engineered foods.

Vitamin B52

Luckily, it wasn't a total downward spiral into nutritional ignorance. Although the science wasn't conclusive yet, people did suspect that processed foods weren't as healthy as the fresh variety, and studies were under way.

An early breakthrough came in 1928 at the University of Wisconsin, where scientists irradiated canned and pasteurized milk with vitamin D, a nutrient it did not naturally have. The effect was soon duplicated with cheese. Many companies, sensing that their ever-growing list of processed foods would eventually come under regulatory scrutiny, began funding vitamin research. At the same time, scientists at the Mayo Clinic in Minnesota were performing vitamin experiments on teenagers. After putting them on a diet low in thiamine, or vitamin B1, researchers found their four subjects became sluggish, moody and "mentally fatigued."[35] They repeated the experiment with six female housekeepers, who found their ability to do chest presses greatly diminished. When two of the six were put on a diet high in thiamine, their abilities recovered.

Russell Wilder, one of the doctors, argued that Hitler was using vitamin deficiency as a weapon in his domination of Europe. The Nazis were "making deliberate use of thiamine starvation to reduce the populations . . . to a state of depression and mental weakness and despair which will make them easier to hold in subjection."[36] Thiamine, Wilder declared, was therefore the "morale vitamin," a vital part of any military effort, not to mention a balanced breakfast.

Thiamine is naturally found in beans, legumes and whole-wheat flour, but in 1940, Americans hated whole-wheat bread— it accounted for only 2 percent of the bread sold.[37] Milling removed between 70 and 80 percent of wheat's thiamine to produce the white bread Americans loved, so Wilder believed some sort of government intervention was needed. Having joined the Council on Foods and Nutrition of the American Medical Association in 1931 and the Committee on Medicine of the National Research Council in 1940, he was already a food authority and in a strong position to proclaim his views on vitamins. In 1941 he organized and became the first chairman of the Food and Nutrition Board of the National Research Council, which put him within earshot of the most powerful American politicians.

In 1942 he finally convinced the government to decree that all flour used by the armed forces and federal institutions should be "enriched," with nutrients such as thiamine mixed back in. The ruling took immediate effect, and by the middle of 1943 about three-quarters of the bread being produced in the United States was enriched with B1.[38] The British military came to the same conclusions. After finding that 41 percent of the young men drafted for service during the First World War were medically unfit, mainly because of poor nutrition, the government also ruled that its flour had to be enriched.[39] The move to counter the bad effects of food-processing technology with *good* food-processing technology had officially begun. Following the war, processors cashed in on the emerging trend toward health consciousness by expanding enrichment practices to other foods, including rice and cereals. They also took it a step further by "fortifying" products, or adding nutrients to foods that did not naturally have them. (The trend went overboard in the fifties, when even chewing gum was imbued with vitamins.)

Enrichment was one step forward to good nutrition, but by the end of the fifties the world had taken a number of steps back. Mass spectrometers, used today by just about everybody—from sports bodies in detecting the use of performance-enhancing drugs to mining companies in finding gold deposits—allowed food processors to alter the chemical make-up of foods. Tastes, textures, shapes and colors could be changed and molded as desired. Canning, dehydration, freezing and drying techniques, as well as packaging made possible by new plastics, all improved the longevity of food, preventing spoilage and allowing for transportation across vast distances. The road was paved for truly international foods and, with them, international food-processing companies.

The wartime boom in refrigerator and freezer sales also continued into the fifties and sixties and, combined with the advent of the microwave oven and plastics, gave households new and easier ways of storing and preparing foods. In the span of three decades, the creation, sale and consumption of food had changed more dramatically than it had over the previous three centuries.

All of this was the product of the emerging prosperity-driven consumer culture. Again, after total war came total living, and food was an integral part of that maxim. Food was no longer the precious commodity it had been during the Great Depression and the wars; time was now at a premium.

Food producers, armed with an arsenal of new technologies, were more than happy to cater to these desires. As with plastics and their eventual environmental harm, when it came to the potential negative health effects of these new processed foods, the collective thinking was, "To hell with them, let's eat."

From Lab to Drive-Thru

Wartime processing advances were only half the food revolution story, though. The second part wasn't taking place in home kitchens and grocery stores, but in restaurants around America. Just as the accelerating pace of post-war life was creating demand for domestic foods that were faster to prepare and easier to store, the same was happening outside the home. Before the war, eating at a restaurant was a rarity for the typical family, but the economic boom led to a wave of new restaurants focused on getting people their meals quickly, cheaply and efficiently. And technological engineering was at their core.

The first fast-food chain to achieve international success, Dairy Queen, was centered on an invention that solved the two biggest problems of the ice cream business: the hardness of the product and the time-consuming manual labor that went into scooping it. The new machine stored ice cream at temperatures just above freezing, so it was cold but soft, and dispensed it with the simple opening of a spigot. This allowed Dairy Queen outlets to pump out hundreds of cones an hour, vastly increasing their volume of business.

Burger King was also founded on a similar volume-enabling machine. The restaurant's contraption cooked four hundred burgers an hour by automatically moving them through a broiler in wire baskets, which was significantly faster than cooking them manually. Kentucky Fried Chicken, meanwhile, used a new type of pressurized deep fryer to cook drumsticks and breasts in one-third the time of conventional deep fryers. Now titans of the industry, all of these chains relied on technology to mass-produce food, achieve ever-increasing sales volumes and drive large-scale expansion.

Of course, none was as successful as McDonald's. In the fast-food industry's early days, no other chain saw as much potential for technology, science and engineering to deliver sales speed and volume, and no one profited as much from investing in innovation. The original McDonald's was started by Richard "Dick" McDonald and his older brother Maurice "Mac," the sons of a shoe factory foreman in New Hampshire. The brothers had moved to California at the height of the Depression in 1930 in search of riches. After trying their hands at managing a movie theater, they opened a drive-in hot dog stand in Pasadena in 1937. The stand was a success but the brothers wanted a higher volume of customers, so they relocated to the nearby boom town of San Bernardino, where they opened the first McDonald's drive-in restaurant in 1940 on busy Route 66. It too was a hit, particularly with teenagers, who used the drive-in as a hangout. By the mid-forties McDonald's, like many of the dozens of other drive-ins dotting California, was raking it in. Together the brothers had created a new style of restaurant, one that was driven by the post-war economic boom, the growing ubiquity of automobiles and an increasing desire for speedy service.

But the operation still wasn't fast enough or achieving the sort of volumes the McDonalds wanted. They were frustrated with the traditional tools and systems of the restaurant business and wanted to apply the assembly-line technology that Henry Ford had used to speed up car manufacturing. The brothers invented their own equipment and "became enamored of any technical improvement that could speed up the work," as one McDonald's historian put it.[40] In the fall of 1948 they closed down the restaurant to refit it purely for speed. They replaced the grill with two bigger custom-built versions and hired a craftsman to design new equipment, much of which is still used in the fast-food business today, such as the broader metal spatulas that allowed for several burgers to be flipped at once, and the handheld stainless steel pump dispensers that squeeze precise amounts of ketchup and mustard onto buns. The McDonalds also purchased four Multimixers, each of which made five milkshakes at a time. The move turned out to be fateful, as it eventually brought them into contact with Ray Kroc.

Besides the gear, the brothers also made major changes to their system. The menu was pared down to just eleven items, and all china and flatware was scrapped in favor of paper bags, wrappers and cups. The carhops, who had served customers at their cars, were fired; customers now had to walk up to the window to place their order. Jobs were regimented into simple tasks, so that two employees were responsible for making milkshakes while another two did nothing but cook fries. The new-and-improved McDonald's, featuring the "Speedee Service System" complete with a sign depicting the cartoon chef "Speedee," re-opened in December as a finely tuned machine, the perfect blend of human and technological efficiency. Sales took a hit initially as the teenagers were scared away, but they soon rebounded and surpassed previous volumes as families discovered the new McDonald's.

By 1954 the brothers were swimming in money and had sold a handful of franchises before their fateful meeting with Kroc. The fifty-three-year-old Chicago native had been a lifelong entrepreneur who had done moderately well, first by selling paper cups and then milkshake machines. His largest customers were using no more than two Multimixers, so he was intrigued by the tiny California operation that had four going at any given time. He traveled to San Bernardino to meet the McDonalds and was awestruck at the lunchtime crowd. Ever the entrepreneur, he knew he had to get on board.

Kroc struck a deal with the brothers for national rights to their restaurant and kicked McDonald's franchising—and its use of technology to build volume—into high gear. Kroc enlisted Jim Schindler to help design his own test franchise in Des Moines, Illinois, near Kroc's Chicago home base. Schindler had trained in electronics in the Army Signal Corps during the Second World War and designed tools for munitions manufacturing, but his most important skills, as far as Kroc was concerned, came from his experience in designing kitchens for submarines. Besides accounting for the cramped environment, submarine kitchens needed to be rugged, easy to clean and standardized, so that one design could be plugged into a variety of ships. Kroc had the same idea for his planned high-volume kitchens, so Schindler, who designed much of the plug-and-play stainless steel equipment found in the growing chain's restaurants, was the perfect fit.

As Kroc's franchising juggernaut gained steam—over four hundred restaurants by 1963—so too did its quality-control problems. McDonald's quickly found it difficult to serve burgers and fries that tasted the same in Los Angeles as they did in New York, and year-round to boot. Kroc turned to the same kind of food science being employed by big processors such as Hormel, Carnation and Nestlé, and in 1957 McDonald's became the first fast-food company to open a research lab. One of the lab's first tasks was perfecting the french fry. Converting to Simplot's frozen potatoes was only one step in the quest. The McScientists also experimented with curing potatoes so that their sugars were converted into starches, and studied the spuds' solids content with new, complex machinery. They found that only potatoes with a solids content of at least 21 percent were acceptable, so they equipped suppliers with hydrometers, devices that measured a potato's gravity and thus its solids content. Few potato farmers had ever seen a hydrometer before, let alone used one, but if they wanted to supply the growing food giant, they had to incorporate the new technology. McDonald's even invented the "potato computer," which was really just a sensor that detected when the oil in the fryer hit the correct temperature. The sensor, which beeped when the fries were perfectly cooked, was later used with all fried products, including Chicken McNuggets and Filet-o-Fish, and is now standard across the industry. McDonald's quest for the perfect french fry, an endeavor that cost the company an estimated $3 million in its first decade, was not unlike unlocking the secrets of the atom.[41]

The company put the same amount of scientific and engineering effort into ensuring the quality and uniformity of its other food offerings. To speed up milkshake production and save space in restaurants, Kroc introduced a new machine that mixed a concentrated liquid mix rather than a frozen ice milk base. The liquid mix was poured into the machine, frozen and automatically dispensed, much like Dairy Queen's soft-serve ice cream. The new system was faster than the old, which required making milkshakes manually, and the cans the liquid mix came in—similar to concentrated orange juice cans—took up less storage space than the frozen ice milk.

Before McDonald's, hamburger was the meat industry's poor cousin, a veritable repository of meat packers' unwanted parts. That didn't suit Kroc, who not only wanted reliability from his burgers, but also preferred to avoid the sort of large-scale lawsuits that would follow if he served people mystery meat. So the company's labs dissected and experimented on ground beef as if it were an alien from another planet. The result was a fifty-item checklist that was passed on to franchisees, who could then test the meat they received from suppliers to see if any shortcuts had been taken.

By the late sixties, McDonald's restaurants received hamburger shipments three times a week from a variety of the company's 175 suppliers. With that much movement coming from so many sources, it would have been easy for some shipments to be overlooked for inspection or for some suppliers to cut corners, a fact that deeply frightened Ray Kroc. "I'd wake up in the middle of the night from dreaming that we had bad beef and thousands of customers with upset stomachs. I wondered how the hell we'd get over something like that."[42]

The solution was to shift the chain's focus on technological innovation from the store level up to the supplier level. In the late sixties, a trio of small meat suppliers formed a small company called Equity Meat with the goal of solving McDonald's beef problem. They came up with a process that was similar to freezing fries—the beef patties were flash frozen cryogenically to below two hundred degrees, which sealed in their juices. The group experimented with different coolants, freezing speeds, meat-grinding techniques and cooking times, and even came up with the first computerized system for ensuring that the right blend of meat went into the burger. While McDonald's executives scoffed at using frozen patties initially, much as they had with frozen fries, after testing they found that the burgers tasted just as good if not better than fresh ones and shrank less when cooked on the grill.

All of a sudden, the company's quality-control problems with meat were solved. By 1973, just two years after testing Equity's burgers, almost every McDonald's restaurant had converted to frozen patties. As per its agreement with McDonald's, Philadelphia-based Equity was required to share its technology with any other supplier the chain saw fit. Four other meat packers were chosen and in one fell swoop McDonald's pared its beef suppliers down from 175 to 5, creating huge efficiencies by consolidating its quality-control management to only a few points. The company was thoroughly grateful to Equity and gave it a good chunk of its burger business, rapidly transforming the small company, now known as Keystone Foods, into one of the biggest beef suppliers in the world. Today, Keystone employs more than 13,000 people and services 30,000 restaurants worldwide, handling 388 million pounds of beef and 1.6 billion pounds of poultry products a year.[43] The other beef suppliers who got Equity's technology, including Golden State and Otto & Sons, also became industry giants in their own right.

McDonald's ushered in a similar transformation of the chicken business. In the late seventies, after realizing there was growing customer demand for poultry, the company hired Rene Arend, an acclaimed chef from Luxembourg, to come up with a chicken product. Arend actually happened upon the chicken McNugget by accident. His Onion Nuggets, which were battered and deep-fried chunks of onion, failed to catch on, but McDonald's CEO Fred Turner suggested he try the idea out with chicken. Arend cut some chicken into bite-size pieces, then battered and fried them up. Turner loved them, but both he and Arend knew that mass-processing chicken in the numbers McDonald's needed was difficult because there was no automated way of removing the bones. KFC might have been selling chicken by the bucket-load, but the rival fast-food chain wasn't chopping its poultry up into the little bits required by McDonald's, which meant the burger chain would have to innovate once more.

Keystone again came to the rescue by modifying a hamburger patty machine so that it cut deboned chicken into the now-familiar McNugget shapes. The poultry still had to be deboned by hand, but the invention greatly sped up production lines. Chicken McNuggets entered test stores in 1980 and were an immediate hit, rolling out company-wide a year later. Keystone shared its technology again, this time with Tyson Foods, which was already one of the largest chicken producers in the world.

Tyson introduced its own improvement to the McNugget process by creating an entirely new breed of chicken. "Mr. McDonald," as the breed was dubbed, was nearly twice as large as traditional broiler chickens; its size didn't just mean more meat, it also made the deboning process much easier. Within three years of their introduction McNuggets accounted for 7.5 percent of McDonald's U.S. sales, and they brought in more than $700 million in 1985, making them one of the most successful products in the company's industry.[44] McDonald's is now the second-largest purchaser of chicken in the world, after KFC.

The poultry industry jumped on the bandwagon and reaped the benefits. The McNugget contract helped Tyson surpass rival ConAgra as the largest U.S. chicken producer in 1986. Deboned chicken, once the ugly duckling of the poultry business, took off and others piled on. KFC jumped on board with its own nuggets a few years after McDonald's. Tyson rivals, meanwhile, responded to this shift in the market by growing their own bigger chickens, a trend that continues to this day as suppliers try to get more bang for their buck with ever-larger birds. As Bud Sweeney, one of the developers responsible for coming up with McDonald's deboning process, put it back in the eighties: "We absolutely revolutionized the chicken business."[45] The trend has angered animal rights activists, though, who are horrified by images of the bloated chickens, unable to support their own weight.

The technological innovations, however, helped McDonald's build big sales volumes and profits, which attracted franchisees eager to make a fortune. McDonald's easily outpaced its competitors in the early days and is now the biggest fast-food operation in the world, with more than 32,000 locations.[46] While Kroc is often praised for his business acumen and his ability to sell and manage franchises, he rarely gets credit for his recognition and use of technology. Without all of the advances Kroc pioneered—he even had Schindler build the world's first rooftop heating and air-conditioning unit—McDonald's may never have grown past the handful of restaurants set up by its founders. Moreover, Kroc's use of technology did more than just drive McDonald's growth, it created a measuring stick for the rest of the industry. While some chains understood that innovation was the chief driver of sales volume, some did not. Those that fell in line prospered while those that failed to grasp the concept either dropped off or remained small. Kroc established the bottom line in the fast-food industry: innovate or die.

McDonald's also revolutionized supply logistics by pioneering the "just-in-time delivery" model that would later be attached to companies such as Wal-Mart and Dell Computer. In 1970 a typical McDonald's restaurant received about twenty-five shipments a week from 200 different suppliers, ranging from burgers and potatoes to paper cups and straws. As a result, restaurants were becoming mini-warehouses. The company responded by encouraging suppliers to consolidate and become "one-stop shops" that could provide all of its needs, from foods to dry goods. These consolidated companies could then set up their own regional distribution centers and storage warehouses, which would in turn supply all of the nearby McDonald's restaurants whenever supplies ran short. McDonald's was thereby able to cut its network of distributors down to about ten companies, and by the mid-eighties, shipments to restaurants were down to about two a week. The new system meant more efficiency, less paperwork, tighter control over purchasing, less waste and fresher food.[47] Virtually the entire food production system in the United States subsequently followed this model.

Critics have raised concerns over this McCentralization of distribution. With large suppliers responsible for entire geographic regions, an outbreak of contaminated food is likely to be more widespread than it would be if smaller companies were dealing with smaller territories. While those arguments have proven to be partially true, McDonald's has carefully preserved competition between its suppliers by steadfastly demanding improvements in quality, and it has not been shy in switching when it catches a supplier cutting corners. The company didn't hesitate to move its sauce-production business from Conway to Golden State, for example, after quality complaints began to mount. The threat of losing a large McContract, which often accounts for much of any supplier's business, has usually been enough to keep them honest. As for contamination issues, while large-scale outbreaks certainly do happen, many food scientists believe such events would be far more frequent, albeit smaller in scale, without the centralized food system McDonald's helped usher in. The proof may very well be in the pudding, or meat and produce—anyone who has taken a vacation in a developing country, where food is grown locally without much inspection, knows how easy it is to get food poisoning. While such infections are usually not serious, they can certainly turn a person off eating.

By the sixties, the food revolution was complete. In the grocery store, consumers could choose from a plethora of ready-to-eat meals—instant soup, TV dinners, frozen fries—that they could heat up with their microwave ovens and store as leftovers in their Tupperware containers. Outside the home, families could drive to the local McDonald's, Burger King or Kentucky Fried Chicken and eat without spending a fortune or taking a long time. Both scenarios gave consumers more time and money to spend on the things they really wanted to do, whether it was hobbies, travel or other leisure activities. With the biggest real necessity—eating—taken care of, consumers were free to indulge themselves in ways they never had before. Is it a surprise that the sexual revolution soon followed?

ARMING THE AMATEURS

The Bell & Howell Filmo 16-millimetre greatly simplified movie making during the Second World War and into the 1950s.

PHOTO COURTESY OF THE AUTHOR.

War is, after all, the universal perversion. We are all tainted: if we cannot
experience our perversion at first hand we spend our time reading war stories,
the pornography of war; or seeing war films, the blue films of war; or titillating
our senses with the imagination of great deeds, the masturbation of war.

—JOHN RAE, *THE CUSTARD BOYS*

The most surprising by-product of the Second World War has to be the creation of the modern pornography industry. That may sound like an odd statement, so an explanation is in order. The war brought technological standardization to smaller film cameras and trained a horde of amateurs in their use. After the war, these budding filmmakers, shunned by Hollywood as outsiders, took their new-found technology and skills and made their own movies. Many of them were sucked in by the sexual revolution that was unfolding across all media, and a proper industry for pornographic movies was born. The sexual revolution and the film revolution unfolded in tandem, each influencing and furthering the other.

Before the war, the market for films featuring sex and nudity, known as "blue movies" in Europe, was a cottage industry that catered mainly to brothels.[1] Movies were shot and shipped to private buyers, mostly in England and France, but also to distant places such as Russia and the Balkan countries. As a 1967 *Playboy* article pointed out, "By the end of *la belle époque*, no self-respecting brothel on the Continent considered its facilities complete without a stock of these films."[2] The movies were also made in the United States, where the main audience from the twenties through the forties was men at stag parties. Social groups such as the Legionnaires, Shriners, Elks and college fraternities held screenings by quietly renting projectors and a few reels of "stag films" from camera stores in the trade.[3] Both American and European markets for the films hit their peaks in the years before the Second World War as newly introduced 16-millimeter cameras lowered the costs of producing the movies.[4] Still, the market was small—*Playboy* estimated that by 1970, only about 2,000 stag films had been made.[5] Up until the war, film production was haphazard, low-profit and high-risk. In many countries the movies were also illegal, barely tolerated by authorities.

Film historians generally regard France as the birthplace of stag films since the country had the early lead in movie projection with Louis Lumière's *cinématographe*, which made its debut in 1895. But evidence of early pornographic movies has been found virtually everywhere around the globe, including Germany, Russia, Argentina and Japan. The movies—black-and-white, silent and usually ten to fifteen minutes long— were typically shown only in the regions where they were made because producers feared being prosecuted for distributing obscene materials. They featured amateur participants in a variety of settings and were surprisingly graphic for the time, often showing hard-core penetration, gay sex and even bestiality. *A Free Ride*, thought to have been made around 1915 and generally considered to be the first American pornographic movie, featured a man and two women going for a naughty car ride in the country. When the man stopped the car to pee, the women spied on him and became aroused. The man, in turn, spied on them and made sexual advances, culminating in three-way sex.

Despite its carnal content, A Free Ride and many other stag films of the time had surprisingly high production values, suggesting that professional moviemakers were secretly behind them. As film historians Al Di Lauro and Gerald Rabkin noted in their book Dirty Movies, "This professionalism was understandable in an era in which film equipment was expensive and cumbersome, hence unavailable to the amateur."[6]

That changed after the war, when film technology and techniques underwent a boot camp of sorts that greatly decentralized moviemaking from Hollywood and opened the industry for amateurs. The military had a great need for films, which served three key wartime purposes: they recorded enemy forces and weaknesses, helped train soldiers and served as morale-boosting propaganda for viewers back home. Cameras were "as necessary as radar," as one film historian put it.[7]

Hollywood was drafted en masse—fully one-sixth of workers, or 40,000 people, in the production, distribution and exhibition of motion pictures ended up in the armed services. That figure included directors, writers, camera operators, electricians, technicians and machinists.[8] Actors enlisted, Hollywood professionals helped train camera operators for the Army Signal Corps, and feature-film directors including William Wyler, John Huston and Frank Capra shot combat footage and training movies. As with every other industry, Hollywood saw helping the war effort as its patriotic duty.

But while the Signal Corps welcomed Hollywood's expertise and manpower, it didn't necessarily want the industry's technology. At the time, feature films were shot using

35-millimeter cameras—big, hulking monstrosities that rolled on wheels if they moved at all. The equipment was impractical for war conditions, where compactness and mobility were absolute necessities. Hollywood studios also needed to shrink their large film crews into small, mobile, flexible units that could work in the field. The adjustments weren't easy. "The technical requirements of such a small production unit are rather confusing to those accustomed to Hollywood standards," said the American Society of Cinematographers at the time. "The first requirement is absolute mobility, to travel cheaply and quickly under any conditions—plane, auto, boat . . . Equipment must be reduced to a minimum."[9]

Sixteen-millimeter cameras had hit the market back in the early twenties, but they were shunned by studios, who felt they produced inferior pictures. The smaller cameras also packed in less film than their larger 35-millimeter relatives, which relegated them to shooting shorter-length footage such as newsreels. In 1925 Bell & Howell designed the Eyemo, a rugged 35-millimeter camera that was a fraction of the size of others like it, but it too saw limited use by studios because of its short film capacity. Lastly, 16-millimeter cameras were usually hand-cranked, which made them impractical for use on sets. Manufacturers such as Eastman Kodak and Bell & Howell therefore marketed their respective Cine-Kodak and Filmo smaller-gauge cameras to amateur users, but sales were poor since the gadgets were still relatively expensive. In 1932 Kodak developed an even smaller and cheaper 8-millimeter camera for the amateur market, but it too sold modestly because shooting film was still too complicated for the average person.

Sixteen-millimeter film cameras were, however, perfect for military purposes. Front-line units trained by the Army Signal Corps were armed not only with guns but Filmo and Cine-Kodak cameras to capture combat footage for later study. Eyemos were modified to sit on machine-gun tripods while Filmos were equipped with gun-stock attachments so that they could be aimed in much the same manner as rifles. B-17s were equipped with cameras to film bombing runs and fighter planes used them to shoot dogfights. The navy was well equipped to screen training films, where gunners used actual footage to prepare for battle. Some navy ships were nicknamed "floating studios" because of their processing and production facilities.[10]

Manufacturers repurposed their products and ramped up production to meet the sudden massive demand. While Cine-Kodaks were used to shoot family get-togethers during the twenties and thirties, wartime ads from the company proclaimed them "the fighting Cine-Kodaks." Bell & Howell, meanwhile, touted its Eyemo camera as "the air corps super snooper."[11] Camera makers also stepped up their innovation and development to meet the military's needs. In 1943 Bell Telephone Laboratories produced the Fastax, capable of shooting eight thousand frames per second on either 8- or 16-millimeter film, to document the "split-second action of our high-speed war machine."[12] Since cameras broke down easily in war conditions, the standardization of parts became critical. In 1943 the industry, along with the Army Signal Corps, Army Air Force, Army Engineer Corps, the Navy and the Marine Corps, formed the War Standards Committee to oversee camera production. As one historian noted, "Experienced motion picture engineers who formerly solved Hollywood technical problems now answered military needs."[13] The standardization and interchangeability of parts ushered in by the committee helped bring manufacturing costs down, which introduced savings for the buyer.

The military also faced a shortage of people to produce war footage. Experienced Hollywood hands were generally too old to draft, so the task of training new camera operators and technicians fell to the army. In 1941 the Signal Corps opened the Training Film Production Laboratory in New Jersey to train recruits in movie filming and editing, effectively teaching them to shoot cameras as well as guns. Hollywood professionals were brought in to teach intensive six-week courses, even though they were more skilled at filming interior scenes than the exterior shoots that would become the norm during the war. The courses marked the first time the American Society of Cinematographers offered formal training of any kind. Like the equipment itself, soldiers' skills had to be standardized and interchangeable. The result was the mass production of semi-professional moviemakers. "Today, a motion picture is an integral part of every military unit," the ASC reported at the time. "This country has a great untapped reserve of capable cinematographic talent among amateurs and semi-professionals— men who though they may not have made a career out of photography, have yet attained great skill with their 16-millimeter and 8-millimeter cameras."[14]

The Signal Corps was at its peak in the fall of 1944 when more than 350,000 officers and men served, six times as many as in the First World War. The Corps trained more than 432,000 soldiers over the course of the war and more than 34,000 officers graduated from fifty courses. By the time the war ended, the Signal Corps had produced more than 1,300 films.[15] In 1945 the military capitalized on all that footage by setting up the

Academy War Film Library, which rented and sold original war reels to movie studios.

With the conflict over, the stage was set for a revolution in moviemaking. Better mobile cameras meant that location-based shooting of feature films became much more commonplace. Film-making became decentralized and freed from the confines of studio lots. The war had also matured and standardized film technology, which meant that, as of the late forties, cameras intended for the amateur market were actually affordable to those users. Sixteen- and 8-millimeter cameras were much simpler because they all worked in more or less the same way and used the same parts. Moreover, a generation of war veterans returned to their normal lives as semi-professional, trained camera operators. They had not been trained to shoot movies with big, expensive equipment on large studio sets, but rather with small cameras in on-the-go situations. Many of these veterans parlayed their new skills into simple home moviemaking (shooting backyard barbecues, children's birthday parties or family vacations), while others sought to go professional.

The amateur camera equipment market boomed and manufacturers reaped the benefits. Bell & Howell, for one, saw its sales increase by 125 percent between 1947 and 1956, with profit tripling over the same time frame. By 1956 professional motion picture equipment accounted for only 3 percent of the company's sales while amateur gear had risen to 27 percent (military, industrial and educational products made up the bulk of the rest).[16] The lucrative market attracted foreign competitors such as Bolex from Switzerland and sales swelled even more. By 1961 the photographic leisure market was valued at $700 million per year, driven largely by the amateur segment, which more than doubled from 1950 to 1958. Eight-millimeter camera sales also boomed, more than tripling over the same period.[17]

Some of this explosive growth was directly driven by amateurs creating their own stag films. A core of small businesses emerged to produce and sell these films through mail order, while camera stores quietly stocked them for rent and purchase.[18] By 1960 the market had shifted to the point where consumers bought more stag films than they rented.[19] A United States government inquest into pornography in the late sixties found that stag films sold at camera stores also "served as a catalyst for the rental or purchase of movie projectors, screens, cameras and other equipment."[20]

All the Boobs That Are Fit to Print

The explosion in the amateur camera market coincided with the beginnings of the sexual revolution. The groundwork was laid by Dr. Alfred Kinsey, a zoologist at Indiana University, with the publication of *Sexual Behavior in the Human Male* (1948), followed by *Sexual Behavior in the Human Female* (1953). The reports were hugely controversial because they touted statistics that challenged the contemporary, mostly puritanical views regarding sex. Among the most sensational assertions made by the reports were that there were gradations of homosexuality (i.e., you can, in fact, be only partly gay), that more than half of all males cheated on their wives and that nearly a quarter of all men found sadomasochism a turn-on.

The reports were a huge influence on Hugh Hefner, a Chicago native and former copywriter for *Esquire* magazine. Hefner, who had grown up in a puritanical family and harbored an ambitious desire to redefine sexuality, found great inspiration and affirmation in Kinsey's reports. The budding journalist was overjoyed to find that he was not alone in possessing a voracious and non-traditional sexual appetite, and that perhaps such drives were commonplace.

In 1953 Hefner launched *Playboy*, a magazine ostensibly intended to act as a guide to the evolving post-war consumerist culture, but which really hinged on the nude pictures of women it featured. Hefner originally wanted to call his magazine "Stag Party," which would probably have been a more honest title, but he was forced to change it after an outdoors publication called *Stag* objected. (Its editors obviously didn't want any confusion between a magazine that featured nude women and one that featured nude deer.) Hefner was himself a connoisseur of stag films and hosted viewing parties in his Chicago apartment, where he would try to vent any embarrassment among guests by adding funny quips to his narration. He had even made his own stag film, *After the Masquerade*, in which he had sex with a female acquaintance while the two wore masks.[21]

The first issue of *Playboy*, which Hefner had not put his name on for fear of obscenity charges, sold 54,000 copies, or 80 percent of its print run, an astonishing feat for a new magazine. *Playboy*'s circulation rose dramatically over the next two years, hitting half a million by the end of 1955 and one million in 1956, at which point it surpassed Hefner's old employer, the twenty-year-old *Esquire*.[22]

Playboy's success mirrored the rise of consumerism—both were the product of a prolonged period of public denial. Consumers had suffered long and hard, first through the Great Depression, then through the war. As the good times came, so too did conspicuous consumption of all manner of goods: cars, ovens, fridges, television, stereos, plastic goods, toys, processed foods, alcohol and, of course, film equipment. Sex had also been repressed and some of those traditional puritanical forces targeted Hefner, only to be rebuffed by authorities who backed the emerging era of permissiveness. In 1955 Hefner sued the U.S. Post Office for denying *Playboy* a mailing permit and won a total victory. The federal court issued an injunction preventing the Post Office from interfering with the magazine's distribution and awarded *Playboy* $100,000 in damages. Hefner was effectively given the green light to continue his quest to redefine sexuality.

Kinsey and Hefner paved the way for moviemakers to extend the new sexual liberation to the silver screen. The authorities' implicit go-ahead to *Playboy* was extended to director Russ Meyer, whose film *The Immoral Mr. Teas* (1959) became the first soft-core pornographic movie to pass censors and make it to mainstream theaters. Meyer had spent his childhood in Oakland, California, playing with cameras. His parents had divorced in 1922, shortly after he was born, and his mother pawned her wedding ring to buy Russ an early 8-millimeter camera for his fourteenth birthday. By the time the United States joined the war in 1941, Meyer was nineteen and already a self-taught cinematographer. He enlisted and was assigned as a combat photographer to the 166th Signal Photo Company, the photo unit of General George Patton's Third Army, where he served alongside New York native Stanley Kramer. After the war, the two men took distinctly different career paths: Kramer went on to direct Oscar-nominated films such as *Ship of Fools* (1965) and *Guess Who's Coming to Dinner* (1967) while Meyer shot pictorials for Hefner's *Playboy*, then redefined what was permissible in mainstream movies with *The Immoral Mr. Teas*.

The film, which starred Meyer's army buddy Bill Teas as a door-to-door salesman who had the uncanny knack of running into nude, buxom women, was made for only $24,000 but reaped more than $1 million through independent cinemas.[23] As *Time* magazine noted, the movie's success "opened up the floodgates of permissiveness as we know it in these United States."[24] It also kicked off a soft-core pornographic movie gold rush. Meyer, for his part, gave credit where credit was due. "The real driving force behind the increasing permissiveness of our society is Hugh Hefner. I simply put his illustrations to movement."[25]

Attack of the Killer B's

Meyer was a direct product of the post-war boom in amateur film that spawned an entirely new category of alternative, semi-professional productions: the "B" movie. While the term was initially used in the twenties and thirties to identify bonus attractions tacked on to feature films, in the fifties it became synonymous with low-budget releases made by Hollywood outsiders, many of whom were military trained or used war-seasoned technology. The first genres to gain popularity were science fiction and horror, with films such as *Creature from the Black Lagoon* (1954), *Invasion of the Body Snatchers* (1956) and *The Curse of Frankenstein* (1957). Heading into the sixties, B movies splintered into sub-genres that were collectively known as exploitation films, since they often "exploited" lurid subject matter, particularly sex and violence. "Grindhouse" cinemas, so named because the first ones were housed in former "bump-and-grind" burlesque theaters, sprang up in urban areas to accommodate the growing number of exploitation films.

Meyer carved out his own sub-genre of exploitation films—"sexploitation," essentially soft-core pornography—through a string of follow-up hits to *The Immoral Mr. Teas* that included *Wild Gals of the Naked West* (1962), *Faster Pussycat! Kill! Kill!* (1965) and *Vixen!* (1968). At one point, he had four films in *Variety* magazine's top 100 grossing movies of all time.[26] Grindhouse cinemas followed the trend and many converted to showing only adult movies. Before Meyer's breakthrough, only about 60 theaters in the United States showed films that were considered sexploitation. By 1970 that number had climbed to 750, helped largely by the establishment of the Pussycat theater chain.[27]

By the sixties, mainstream Hollywood was feeling the influence of B movies in general and sexploitation movies in particular; the studios were watching how Meyer and the copycat directors he spawned were experimenting with smaller camera technology. "Even Walt Disney was curiously interested in what would happen," Meyer said.[28] They were also watching the steadily increasing box-office returns: *Vixen,* which had been made for just $72,000, raked in an astonishing $15 million.[29]

The studios moved to attract this growing audience with more sexually daring mainstream movies, culminating in 1969 with *Easy Rider* and *Midnight Cowboy*. Meyer also made the full jump to the mainstream when 20th Century Fox picked up his *Beyond the Valley of the Dolls* (1972) for distribution. With B-movie sexploitation directors and Hollywood studios depicting increasingly graphic sex, the stage was set for a film that pushed the envelope by bringing the hard-core action of stag films to the mainstream.

That film was *Deep Throat* (1972), the story of a sexually frustrated woman who is unable to achieve an orgasm until she discovers that her clitoris is in her throat. *Deep Throat* featured explicit oral and anal sex and was given an X rating by the Motion Picture Association of America. The hour-long film also produced massive box-office numbers. Shot over six days in Miami at a cost of only $25,000, it ended up grossing an estimated $100 million in combined box-office and video receipts.[30] *Deep Throat* established the porno formula of low-budget, star-driven films that has been followed ever since.

Sexploitation and the mainstream porno blitz didn't kill off stag films, however. In fact, short, amateur-produced films found an entirely new and bigger market in the early seventies when two self-styled "kings of pornography" teamed up to bring peep shows to the United States. Reuben Sturman, a Cleveland native and Second World War veteran of the Army Air Corps, had started out in the fifties selling comic books from his car. He expanded his business until he was a major magazine wholesaler with operations in several American cities. In the early sixties he moved into the burgeoning field of adult magazines, riding *Playboy*'s success, and by the end of the decade he was the biggest distributor of such publications to a growing number of specialty stores in the United States.

Algerian-born Italian Lasse Braun, a law school graduate living in Sweden, was building his Stockholm-based sex empire at the same time. Braun had filmed his first short sex movie, *Golden Butterfly*, in 1965 using Kodak's new Super 8 camera, an improved version of the traditional 8-millimeter. The ten-minute color film starred Braun himself as a naval officer and a girlfriend dressed up as a Japanese geisha girl. As the director put it, the characters in his film "made love in her boudoir without hiding anything of what makes sex so luscious and irresistible."[31]

Having already established some notoriety in Europe by publishing magazines and writing erotic novels, Braun formed AB Beta Film in 1966 to produce porno movies. In 1971, with Braun as his primary film supplier, Sturman brought a new product—the peep booth, which consisted of a coin-operated 16-millimeter projector, a small screen and a lockable door—to his adult bookstore customers. Braun linked his stag films together into longer loops, which customers watched for 25 cents per thirty-second to two-minute pop. The booths were extraordinarily successful and transformed the business of stag films. Not only were they rebranded as "loops," the films also shed their regional boundaries through Sturman's nation-wide distribution network. Sturman became exceptionally wealthy, raking in an estimated $2 billion in the seventies. By the late eighties, his personal fortune was estimated at between $200 million and $300 million while his magazine and video empire grossed about $1 million a day.[32]

Film historians refer to the sixties and seventies as the golden age of pornography, a time when seemingly anything was allowed at the movies. But this golden age brewed powerful reactionary forces. In 1970 a Congressional panel set up by Democratic president Lyndon B. Johnson to study pornography's social effects made some fairly liberal recommendations: that children should receive sex education and that no further restrictions should be placed on entertainment intended for adults. Incoming Republican president Richard Nixon, however, gave the report a hostile review, denouncing Johnson's panel as "morally bankrupt" and vowing to fight pornography in all its forms. "So long as I am in the White House, there will be no relaxation of the national effort to control and eliminate smut from our national life," he said. "Pornography can corrupt a society and civilization . . . The warped and brutal portrayal of sex in books, plays, magazines and movies, if not halted and reversed, could poison the wellsprings of American and Western culture and civilization."[33]

As the eighties neared, the golden age of pornography faced a growing conservative enemy. The amateurs who had ushered in the era armed with war-seasoned technology were in for the biggest fight of their lives. Sturman, whom the U.S. Justice Department believed to be the largest distributor of hard-core pornography in the country, proved to be one of the biggest casualties. In 1989 authorities nailed him for failing to pay $29 million in taxes. Sturman was sentenced to hard time; he managed to escape from his California cell but was recaptured eight weeks later and ended up dying in a Kentucky prison in 1997.

The genie was out of the bottle, however. In less than thirty years, a large group of military-trained semi-professional filmmakers armed with new, high-tech cameras had lured pornographic movies out of insular men's clubs and into downtown theaters in every major city. The same advances that produced Oscar-winning filmmakers such as Stanley Kramer also produced Russ Meyer, Deep Throat and the rest of the pornographic film business. And even better technology— the VCR and later, the Internet—that would further lower the cost of filming and disseminating sex movies was on the way. Social attitudes toward sex in the media had also shifted away from conservatism and toward permissiveness. Just as men and women were experimenting with the new-found freedoms brought on by the sexual revolution and its scientific advances like the birth control pill, they were also becoming more open to how they were entertained. To the dismay of Nixon and his fellow conservatives, pornography was here to stay.

Doing It Yourself

One of the reasons people made such naughty uses of film cameras is that they had been primed for decades with still photos. Before the forties, taking your own nude pictures was a risky proposition. If you tried to develop the film, chances were good you'd run afoul of obscenity laws and end up in jail. The only solution was to either invest in a darkroom or find a shop that offered discreet processing of such "special" photos. Luckily, two pieces of technology came along to cater to this need.

In 1932 American chemist Edwin Land founded a company based on a polarizing filter he had invented, which he used to make sunglasses and camera lens attachments. During the Second World War, Land's company, which he named Polaroid in 1937, supplied Allied military forces with goggles, target finders and other optical equipment. His big breakthrough came in 1948 with the Land Camera, a device that could take a photo and instantly develop it, giving the picture taker a print within minutes. The device and its follow-up models proved to be huge hits. For the average consumer, the cameras—known simply as Polaroids—were the first easy way to take photos, without the additional expense and hassle of waiting for developing and processing. They also provided an easy and *safe* way to create homemade sexual content.

Consumers were, as porn historian Jonathan Coopersmith puts it, finally free from the "censoring eye of the local druggist or the ogling leer of the film laboratory technician."[34] More to the point, Polaroid "was an enormous breakthrough for the amateur because anybody could shoot whatever they wanted," according to *Playboy*'s current photo director Gary Cole. "If you shot pictures of your girlfriend, you didn't have to worry about anybody else seeing them."[35] (The cameras also proved invaluable to professionals, like *Playboy* photographers. With nude shoots sometimes taking days to meet Hugh Hefner's exacting standards, Polaroids helped photographers set up shots and save time and money. "A lot of the initial lighting and posing changes were made at the Polaroid stages before we shot the actual film," Cole says. "It would save you a whole half day, shooting Polaroid.")

Land's company knew about the use of its cameras for sexual purposes, both by amateurs and professionals, and tacitly acknowledged as much. In 1966 the company released a new model called "the Swinger," a term coined in the late fifties to describe a sexually liberated person. (By the seventies, the word had evolved to mean a person who was promiscuous and/or liked to swap sexual partners.) Television commercials for the camera showed good-looking, nearly nude couples frolicking on the beach, taking pictures of each other. With the sexual revolution in full swing, Polaroid was clearly looking to cash in on the new wave of liberation.

The same held true for photo booths, which were essentially giant, immobile Polaroid cameras. By the thirties, fully self-sufficient booths were popping up in public places such as amusement parks and stores. By the fifties, proprietors were encountering an unexpected problem. According to one historian, "Complaints started coming in from Woolworth's and other stores that people, particularly women, were stripping off their clothes for the private photo booth camera. Couples started being a little more adventurous in the privacy of the curtained booth." As a result, many Woolworth's stores removed their curtains to discourage "naughty encounters."[36]

There just seemed to be something about the booths that brought out the exhibitionist in people, a phenomenon that continues to hold true. Brett Ratner, director of such mainstream movies as *Rush Hour* and *Red Dragon*, installed a black-and-white booth in his house and had his celebrity friends take pictures whenever they stopped by. He published the results in a 2003 book, *Hillhaven Lodge: The Photo Booth Pictures*, with some glaring omissions. "There were a lot of middle fingers, a lot of people with their tongues out," Ratner says. "There was also a lot of flashing . . . I didn't publish those."[37]

Booth manufacturers have always known what buttered their bread, and they quietly encouraged such uses. One American manufacturer, Auto Photo, handed out programs at a mid-fifties industry convention that featured a drawing of a woman exposing herself with the caption, "Make sure he remembers you! Send a foto to your boyfriend."[38] Technology makers were happy to reap the benefits of the public's evolving sexual liberation, but it wasn't something they were willing to tout openly. "The fact that this was in the sales material says, okay, these guys know about it, they just aren't talking about it in public," says Coopersmith, who has written several papers on the role of pornography in technological development. "Even now, many of the firms producing the technology and who benefited do not like to publicly talk about it. You still can't tell people what you're doing that openly."[39]

Participatory porn took a huge step forward with the advent of camcorders in the early eighties. While the 16- and 8-millimeter cameras developed during the Second World War were great for amateur pornographers with an interest and knowledge of film, it wasn't until the camcorder that the mass market finally had a viable, easy-to-use video option. The camcorder made filming idiot-proof—just point and shoot, no need to set lighting levels or focus, and no need to get film developed. While the technology obviously had many non-pornographic uses, science-fiction author Isaac Asimov was right when he stated in 1981 that "the age of home video will fundamentally alter our approach to sex."[40] The camcorder took the do-it-yourself porn started by Polaroids, photo booths and smaller film cameras to the next level. As had happened with stag films, a market for amateur porn movies sprung up in the eighties, with the back pages of every adult magazine full of ads for homemade videos. The trend further evolved with the advent of digitalization, with amateur porn websites springing up in huge numbers.

The continually evolving ability for amateurs to create their own sexual media, coupled with an ever-increasing amount of purchasable porn, has dramatically transformed our views on sex over the past half century. On the supply side, technology manufacturers—who once refused to publicly acknowledge the role sexual content played in the success of their products—are starting to come around, especially smaller companies that are desperate for customers of any sort. Spatial View, a small Toronto-based company that is developing software to view three-dimensional photos on digital devices, is just one example. In early 2009 the company gave the honor of announcing the availability of its new Wazabee 3DeeShell, which slips onto an iPhone and allows the user to view specially coded photos and movies in 3-D, to adult producer Pink Visual. The adult company put out a press release touting itself as one of the first content suppliers for the shell days before Spatial View itself announced the product. "They wanted to show that they were on the cutting edge of technology," says Brad Casemore, Spatial View's vice-president of business development. "They came right after us. They read about it and said, 'We'd like to give it a try.' It's not really a target market for us for a variety of reasons, but when a customer comes to you, there's not much you can do."[41]

A GAME OF WAR

Tennis
for Two,
considered by
many to be
the first real
video
game,
used an
oscilloscope
as its display.
PHOTO COURTESY
OF THE AUTHOR.

The sand of the desert is sodden red,—
Red with the wreck of a square that broke;—
The Gatling's jammed and the Colonel dead,
And the regiment blind with the dust and smoke.
The river of death has brimmed its banks,
And England's far, and Honour a name,
But the voice of a schoolboy rallies the ranks:
"Play up! play up! and play the game!"
—HENRY JOHN NEWBOLT, "VITAÏ LAMPADA"

When I was a kid, playtime consisted of one of two activities. Either I'd be running around outside playing "army" or I'd be indoors acting out battles between G.I. Joe troops and their sworn enemy, the evil terrorist organization Cobra. In the outdoor scenario, my friends and I would form teams and "fight" each other with plastic guns on the forested hillside across the street from my house. When it was too cold to go outside, I'd build Cobra fortresses in my basement out of couch cushions and boxes that the G.I. Joe troops would have to overrun.

What was the result of all this war-themed play? Well, I can safely say I'm a master of weapon sound effects. There isn't a variation of "pow-pow-pow," "budda-budda-budda" or "voooooosh" that I can't mimic. But I wasn't a particularly violent child, nor were my parents warmongers who encouraged such military-oriented leisure. No, I was just like millions of children—mainly boys—who happen to like playing war.

For as long as there have been weapons, children have fashioned their own makeshift versions from sticks and other found materials to play at being soldiers, pirates or policemen. Toy soldiers have been around almost as long; wooden carvings have been found in ancient Egyptian tombs, while tin versions were first manufactured in Europe during the Middle Ages. The first plastic toy soldiers, or "army men," were produced in the United States in the thirties and took off in the fifties after polyethylene became available. Toymaker Hasbro further capitalized on boys' fascination with war-themed toys in 1963 when it introduced G.I. Joe, a foot-tall "action figure" influenced by the Second World War. The line sold millions of units while a newer, three-and-three-quarter-inch iteration of G.I. Joe troops—the ones I loved so much—became the biggest-selling toy line of the eighties.[1]

While this link between war and toys has always existed, it wasn't until after the Second World War that military technology began to drive the development of toys and games. Just as military-refined technology created playthings for adults (like cameras that could be used to shoot sexual escapades), a host of entrepreneurial inventors followed the lead of companies like Raytheon, Hormel and DuPont in exploiting their wartime inventions for post-war commercial success. For some of these entrepreneurs, it was all about money. For others, the motivations were deeper. Turning their inventions into toys and games allowed them to show off their creations publicly, a welcome and sought-after escape from the secretive world they normally worked in. Still others sought to entertain and amuse, perhaps as penance for the horrific deeds that some of their other creations were responsible for. Taken together, their efforts have gradually changed our attitudes toward war. Today, the tide has turned completely—the development of toys and games now drives the military. Remote-control robots incorporate the same controllers used in Playstation and Xbox consoles, while troops familiarize themselves with combat zones by playing specially designed three-dimensional games that use the same technology as the *Call of Duty* and Tom Clancy titles found on the shelves of Wal-Mart. In many ways, technology has turned war into a game.

Springs and Things

It started with simple inventions like the humble Slinky. In 1943 navy engineer Richard James was trying to figure out a way to stabilize sensitive instruments on board ships using springs. While tinkering in his home in Philadelphia, he accidentally knocked a steel torsion spring off a shelf. Rather than falling and landing in a heap, the spring—a coil with no compression or tension—"stepped" down from the shelf to a stack of books, then to a tabletop, then onto the floor, where it recoiled and stood upright. The engineer was even more astonished when he gently pushed the spring down a flight of stairs, only to see it gracefully "walk" down. His wife Betty was equally impressed by the spring's eloquent movements and described them as "slinky," which stuck as a name.

The couple thought they might have a hit toy on their hands so they formed a business, James Spring & Wire Company, and took out a loan, which Richard used to make a machine that wound Slinky units. They shopped the Slinky around to local department stores and found a taker in Gimbels Brothers, which set up a display—complete with an inclined plane to demonstrate the spring's walking ability—in a downtown Philadelphia store during the 1945 Christmas season. The Jameses were flabbergasted when all 400 units, which Richard had spent days winding, quickly sold out.[2] The Slinky sensation was off and running.

The couple built a production factory in 1948 to cope with demand and eventually developed spinoff products such as the smaller Slinky Jr. and the Slinky Dog, plus non-spring toys like building kits. For the next decade, the Jameses watched the riches pour in. In 1960, however, Richard became unwound, so to speak. After suffering a nervous breakdown, he left Betty and their six children to join a religious cult in Bolivia. Betty was left to manage the company, which she renamed James Industries, as well as the large debts incurred by her husband's religious donations. She recovered from the shocking turn of events and eventually took the Slinky to new heights, but it wasn't easy. "He had given so much away that I was almost bankrupt. I sold the factory and decided to move from the Philadelphia area back to Altoona, where I grew up, with the business," she later recalled.[3]

Betty helped create the toy's memorable television ad campaign, which featured the catchy jingle that anyone born before the nineties is unlikely to forget: "It's Slinky, it's Slinky, for fun it's a wonderful toy. It's Slinky, it's Slinky, it's fun for a girl and a boy!" By its sixtieth anniversary in 2005 more than 300 million Slinky toys had been sold.[4] A few years earlier, on November 4, 2001, the General Assembly of Pennsylvania named the Slinky the official state toy of Philadelphia. Richard James, however, didn't get to see his invention honored—he died in Bolivia in 1974. The Slinky, meanwhile, went full circle with its military connection, when American soldiers in the Vietnam War found it could be used as an antenna for their mobile radios.

The Slinky sparked the imaginations of military minds and marketers alike, who realized that war technology might just be an untapped gold mine of toy possibilities. Silly Putty came next.

When the Japanese occupied the rubber-producing islands of the Pacific during the early forties, the Allies found themselves facing a potential shortage of vehicle tires and boot soles. Credit for the invention of Silly Putty is disputed; Earl Warrick, a scientist working for Dow Corning, claimed to have created it, but Crayola, the company that now owns the trademark, considers James Wright the proper inventor. Wright, a Scottish engineer working in General Electric's labs in New Haven, Connecticut, came up with a potential solution to the rubber problem when he mixed boric acid and silicon oil in a test tube. The new substance had rubber-like qualities and a very high melting temperature, but it also bounced, stretched further and resisted mold. The putty-like goo, however, wasn't solid enough to replace rubber, so it sat out the war. GE sent the substance to scientists and engineers around the world after the war, but eventually gave up and declared it had no practical use.[5] (The Allied rubber problem, meanwhile, was solved in the late stages of the war when a number of companies, including tire makers Firestone and Goodrich, came up with a synthetic elastomer under a patent-sharing program overseen by the American government.)

Ultimately, GE's substance didn't have to travel far to find a use. Ruth Fallgetter, a toy store owner in New Haven, got her hands on one of the samples that were circulating and immediately saw its potential as a plaything. She brought in Peter Hodgson, a local marketing consultant, to help sell it.

Children immediately fell in love with the putty because of its pliability and ability to copy pictures and text when pressed against newspaper and comics pages. Packaged in a clear case with a price tag of two dollars, the putty outsold just about everything in the store. But Fallgetter wasn't convinced of its long-term viability, so she left it to her partner to take further. Eyeing the Slinky's success, Hodgson bounced around potential names before trademarking one he thought represented the goo perfectly: Silly Putty. He bought production rights and a batch of the substance from GE, then packaged it in plastic eggs, since Easter was on the way. He introduced Silly Putty to potential distributors at the 1950 International Toy Fair in New York, but once again it flopped—nearly all of the toy marketers at the fair advised Hodgson to give up. The persistent entrepreneur didn't listen, however, and eventually convinced department chain Neiman Marcus and bookseller Doubleday to sell the substance for $1 per egg in their stores. Hodgson was convinced Silly Putty was a great toy because it sparked children's imaginations. "It really has a sort of personality," he later said, "and it reflects your personality. . . . A lot of what makes it work is that the stuff in the egg is only the half of it."[6]

Silly Putty sold modestly until it made an appearance in *The New Yorker*, which quoted a Doubleday employee as saying it was the "most terrific item it has handled since *Forever Amber*," a bestselling 1944 novel.[7] More than a quarter million orders rolled in from stores around the country in the three days following the story's publication.[8]

Hodgson and his goo weren't out of the woods yet. The Korean War and the American government's rationing of silicone, the main ingredient in Silly Putty, almost put him out of business in 1951. He had to scale back production, but when restrictions were lifted a year later, his business went straight to the moon, literally. American sales boomed throughout the fifties and Silly Putty also became a hit in several European countries.

But Hodgson scored his biggest publicity coup yet in 1968, when it was reported that astronauts on the Apollo 8 moon mission were using it to secure tools in zero gravity. The entrepreneur's tenacity had finally paid off and he rode Silly Putty to riches. When he died in 1976 Hodgson left behind an estate worth $140 million. Crayola bought the rights to Silly Putty in 1977 and ten years later was selling more than two million eggs annually.[9]

Rocket Doll

As the sixties approached, toys started to become more complex, and so did the technology behind them. The all-time bestselling toy, the Barbie doll, was the product of space-age military thinking. It also had lascivious if not outright pornographic origins. The doll was inspired by Lilli, a cartoon character from the German tabloid newspaper *Bild Zeitung*. *Bild*, as it is now known, was founded in Hamburg in 1952 for people with poor reading skills. Like many tabloids, it made heavy use of photographs and featured news stories that were often sensationalistic and based on dubious facts. Lilli, a tall, statuesque character with platinum-blond hair created by artist Reinhard Beuthien, fit in well with the newspaper's lowbrow editorial direction. She was unabashedly sexual, a gold digger, an exhibitionist and a floozy with "the body of a Vargas Girl, the brains of Pia Zadora and the morals of Xaviera Hollander," in the words of one Barbie biographer.[10] In her first cartoon, Lilli sat in a fortune teller's tent and, after being told she'd meet a wealthy and good-looking suitor, asked, "Can't you tell me the name and address of this rich and handsome man?" Another exploit found her naked in her female friend's apartment concealing her vital parts with a newspaper, saying, "We had a fight and he took back all the presents he gave me," while yet another had a policeman warn her that her two-piece bathing suit was illegal. "Oh," she replied, "and in your opinion, which part should I take off?"[11]

In 1955, looking to cash in on Lilli's popularity, the newspaper commissioned German toy maker Greiner & Hauser to make an eleven-and-a-half-inch doll, complete with ponytail and removable outfits, aimed at adults. The curvy dolls were dressed in low-cut blouses, stiletto heels, skimpy skirts and shorts, and came bundled with innuendo-laced packaging and marketing. "Whether more or less naked, Lilli is always discreet," read one promotional brochure; as a "mascot for your car," Lilli promises a "swift ride," read another. Her wardrobe made her "the star of every bar."[12] The doll sold well to German men as a gag gift, but Greiner & Hauser eventually toned down its overt sexuality to appeal to the larger children's market.

Ruth Handler first discovered the Lilli doll while on vacation in Europe. The president of American toy maker Mattel had been thinking of introducing an adult doll after seeing her daughter Barbara playing with her own paper cut-out creations. Barbara had not been imagining her paper dolls in childish situations, like playing in a sandbox or skipping rope with friends, but rather in much older scenarios, such as high school, college or an adult career. Handler believed there was a market in going against the conventional wisdom of the time, that young girls were only interested in playing with young dolls. Intrigued by Lilli, Handler bought three: one for herself and two for her daughter.

Mattel had been started in southern California in 1945 by Harold "Matt" Matson and Handler's husband, Elliot; the company name was a contraction of the founders' names.

Matson, however, decided against gambling his life savings on the company and sold out to his partner the following year, a move he surely rued after Mattel became a toy juggernaut. Ruth, a stenographer for Paramount Pictures in Hollywood, quit her job and came on board as president. Unlike Matson, the Handlers were every bit the entrepreneurs—and big believers in plastics and other futuristic technology. During the war, Elliot built furniture out of Plexiglas in his garage, then expanded to plastic jewellery, candleholders and other novelties. Mattel's first hit toy was the Ukedoodle, a plastic ukulele, in 1947, followed in 1955 by its very own toy weapon, the Burp Gun. By the time Ruth discovered Lilli, Mattel was already a modest success, with a net worth of $500,000.[13]

Elliot's passion for futurism was stoked when he was approached in the early fifties by Jack Ryan, a Yale-educated engineer working for Raytheon, with an idea for a toy transistor radio. Handler didn't like the idea, but he was impressed with Ryan's knowledge of transistors and electrical circuitry and believed the engineer had the "space-age savvy" to make his own high-tech fantasies real.[14] The entrepreneurial Ryan, on the other hand, wasn't content with his role at Raytheon, where he was working on a team designing the Hawk surface-to-air missile. Handler wooed Ryan away from Raytheon with a contract that promised royalties on every toy patent he came up with.

Ryan became better known for his colorful and extravagant sex life than his engineering skills.[15] After he achieved success with Mattel, he spared no expense in building a hedonistic pleasure palace in California that would have made Hugh Hefner envious. On a five-acre estate in Bel Air that once belonged to actor George Hamilton, the engineer constructed a cross between a castle and a theme park, a mansion complete with turrets, tapestries, eighteen bathrooms, seven kitchens and a tree house that could seat twelve for dinner. By dialing particular numbers on one of the 150 telephones around the mansion, Ryan could activate a waterfall, light up the tennis court, close the front gate, turn on the stereo system or order caviar for the tree house, which had its own chandelier and panoramic view of Los Angeles.

He threw frequent and lavish parties—182 in one year, or one every two days—complete with jugglers, fortune tellers, handwriting analysts, musicians, go-go dancers, minstrels and harpsichordists. Drunken guests bounced up and down on a trampoline or fed the ducks, geese and pony kept on the grounds. Ryan also housed the twelve UCLA students he had recruited to be part of his Mattel design team and courted numerous high-society and celebrity mistresses.[16] In the basement, he put together a black-painted dungeon decked out in black fox fur, and cuddled with his numerous girlfriends under wolf-fur covers in a guest room bedecked with mirrored walls.

Ryan divorced his first wife, ironically named Barbara, and married actress Zsa Zsa Gabor in 1976. The two never moved in together, however, and Gabor soon discovered Ryan's wild side. While the couple was honeymooning in Japan, Ryan paid their guide to have sex with Gabor while he held business meetings with a toy manufacturer, much to his wife's shock and displeasure. "Jack, my new husband (it now appeared) was a full-blown seventies-style swinger into wife-swapping and sundry sexual pursuits as a way of life," Gabor recalled in her memoirs. It took seven months for Gabor to tire of the fact that both Ryan's ex-wife Barbara and two mistresses were living in his mansion.

The couple ended their marriage as abruptly as it had begun. "Jack's sex life would have made the average *Penthouse* reader blanch with shock," Gabor said. "I wanted no part whatsoever in any of it."[17]

Ryan's intersection with the Lilli doll was the perfect circumstance, a twist of fate that provided him with the fortune he needed to indulge his wild desires. Just as he was leaving for Tokyo in July 1957 to find a manufacturer for some mechanical toys he had designed, Ruth Handler passed him her Lilli doll. "See if you can get this copied," she said.[18] Ryan landed a deal with Kokusai Boeki Kaisha to manufacture the doll, then set about redesigning Lilli so she would look less like a "German streetwalker."[19] Ryan's modifications were both technical and cosmetic. Aside from convincing KBK to use a new rotation-molding process to create a softer doll, he also designed and patented new arm and leg joints that gave Mattel's toy greater flexibility. The expertise he had gained in building miniaturized moving parts for Raytheon's missile projects came in handy when designing the doll's joints, which would have to stand up to the rigors of constant play. The levels of stress that young girls could place on their toys weren't unlike the forces of gravity, velocity and drag that missiles had to deal with.

Ryan also did away with Lilli's pouty lips, heavy eyelashes, widow's peak hairdo and built-in heels. The newly redesigned doll, named Barbie in honor of the Handlers' daughter, was introduced to the American Toy Fair in New York in March 1959. Backed by a major marketing campaign, Barbie-mania exploded. Girls loved the doll's wide assortment of accessories and wardrobe and the poseability offered by Ryan's joint designs. Mattel sold more than 350,000 dolls in the first year of production, a number that mushroomed over the next four decades to more than one billion sold across 150 countries. The Handlers cashed in and took Mattel public in 1960, making it the world's biggest toy maker and a member of the exclusive Fortune 500 club. By the late-2000s, the company proudly boasted that three Barbie dolls were sold every second.[20]

Ryan wasn't finished, however. While Barbie was starting her meteoric rise, he put his transistor expertise to use by creating a talking version of the doll. Again, only an engineer schooled in the miniaturization of weapons design could have pulled it off. The result was Chatty Cathy, a doll that spoke eleven phrases—such as "Tell me a story" and "I love you"—when you pulled a drawstring on her back. The doll, which went on sale in 1960, had to be larger than Barbie because of the miniature phonograph record player secured in its abdomen. The record was driven by a metal coil that was wound when the drawstring was pulled. Chatty Cathy proved popular too and was soon second only to Barbie in sales. The drawstring-activated voice function revolutionized the entire toy market when Mattel incorporated it into later products, including its Bugs Bunny series and the See 'n Say educational line.

The former Raytheon engineer also had a hand in creating Hot Wheels, Mattel's second-most successful toy line overall. In the early sixties, Ryan's design team was searching for a way to duplicate Barbie's success with boys. Elliot Handler found the solution when he discovered one of his grandchildren playing with a Matchbox die-cast toy car made by Britain's Lesney Products. Handler decided Mattel needed to get into the car business and hired Harry Bentley Bradley, a designer who worked for Cadillac, to help Ryan create a line of toy vehicles.

Handler loved Bradley's own real-world car, a customized Chevy El Camino with red striped tires, mag wheels and fuel injector stacks protruding from the hood. He told his new designer to use his own "hot wheels" as the basis of the toy line. Handler's plan had one big problem—Lesney's Matchbox line already had a commanding share of the market. Mattel's marketing department warned him that if he wanted to compete, he would need to differentiate his product somehow. Ryan, Bradley and the rest of the design team came to the rescue.

They found that kids liked to race their cars, but existing toy cars didn't roll very well. Ryan's team designed a bent-axle torsion bar suspension system, essentially a miniaturized version of the one used in real cars, with inner wheel ball bearings made of a plastic called Delrin, synthesized by DuPont in 1952. The outer tires were made of nylon and molded into a slightly conical shape, which reduced friction by limiting the wheels to a single point of surface contact. The resulting toy cars had a little bounce to them, much like real vehicles, and rolled very fast.

The team also perfected "Spectraflame," a new method of painting that coated the cars with a gleaming silver layer of zinc, then covered them with a thin layer of candy-colored hues to reflect the "California custom" look Handler wanted. Mattel's cars were not only faster than their Matchbox counterparts, their snazzy paint jobs stood in stark contrast to the drab enamel shades of their rivals. It was like comparing Ferraris to Edsels.

Even before Hot Wheels made their debut at the New York Toy Fair in 1968, they were a huge hit. Mattel estimated an initial production run of ten to fifteen million cars, but when the toys were showcased to Kmart shortly before the fair, the retail chain immediately ordered fifty million.[21] Hot Wheels joined Barbie and Chatty Cathy as the lynchpins of Mattel's toy empire.

In the sixties, Ryan tried unsuccessfully to poach some old cohorts from Raytheon to join his design team. Raytheon archivist Norman Krim remembers visiting Ryan in California to entertain one of his job offers. He was struck not so much by the discussion of possible employment, but by some of the engineer's talking doll prototypes. "He had a bunch of Barbie dolls with tape recorders [in them] that were all saying very nasty four-letter words," Krim says. "He was a crazy guy."[22]

Atomic Tennis

Mattel's products were a step toward more complex toys, but they were nothing compared to what came after. I visited the site of the next big development on a chilly evening in September 2008. My arrival at the Brookhaven National Laboratory, one of the world's best-regarded science and technology hubs, did not go as I expected. While ostensibly "in New York," the lab is actually far out on the eastern end of Long Island, a two-hour train ride from Manhattan. From the train stop, it's a further $25 cab ride—there is no other public transportation servicing the lab. When I got to the front gate at 10 P.M., it turned out there had been a mix-up over which day I was supposed to arrive. Apparently I was early, officially by two hours, and the stern security guard refused to let me in. I had little choice but to sit shivering on a bench off to the side of a small clearing in the forest and wait for midnight to arrive. Luckily a family of deer popped into the clearing for a late-night snack of grass, which provided a small measure of entertainment. One thought kept bouncing around in my head: "I can't believe this is where video games were invented."

Brookhaven was established on the site of Camp Upton, a wartime army base, by the Atomic Energy Commission in 1947 to continue the nuclear research begun by the Manhattan Project. The base was chosen because of its remote location, which was perfect for the nature of the volatile—and potentially destructive—research that would go on there. My visit coincided with the fiftieth anniversary of Tennis for Two, an invention many computer historians consider to be the first real video game. The lab's public relations department was planning a big media event where a reconstructed version of the game would be shown off side by side with the latest Nintendo Wii tennis game, a juxtaposition to show just how far the technology has come in fifty years.

While Peter Hodgson, Richard James and Jack Ryan all looked to profit from inventions or knowledge developed from military origins, William Higinbotham had a distinctly different outlook in creating Tennis for Two. Higinbotham was a graduate student in physics at Cornell University when the Second World War broke out. Karl Compton, president of MIT and a key member of Vannevar Bush's newly minted National Defense Research Council, wasted no time in recruiting the promising young physicist to his Radiation Lab, where research was under way on what would become radar. While at MIT, Higinbotham likely crossed paths with Percy Spencer and his Raytheon team, who eventually took over mass production of radar.

With most of the pressing research work on the detection system done by the middle of the war, Higinbotham was recruited by J. Robert Oppenheimer to head up the electronics division of the Manhattan Project, where he created the timing circuits for the atomic bomb. Higinbotham witnessed the test firing of the first bomb in the desert near Los Alamos, New Mexico, in a shelter 24 miles from ground zero. Like many of the scientists who worked on the Manhattan Project, he was deeply troubled by the fruits of his labor. After the first test blast, Higinbotham and the other observers got into their transport trucks and rode back to the Los Alamos base in complete silence. No one had anything to say.[23] However, fellow physicist Kenneth Bainbridge, a specialist in mass spectrometers, was considerably more vocal than his colleagues later on. The blast was a "foul and awesome display," he told Oppenheimer, and "now we are all sons of bitches."[24]

Even before the bombs were dropped on Japan, scientists at several American research facilities were organizing into protest groups to speak out against the actual use of the weapons. Independent associations formed at the Metallurgical Lab at the University of Chicago, the Clinton Laboratory at Oak Ridge, the Substitute Alloy Materials Lab at Columbia University and, under Higinbotham's leadership, at Los Alamos. When the war ended and the scattered researchers and engineers were finally allowed to communicate with each other, they banded together to form the Federation of Atomic Scientists, a lobby group that sought to limit proliferation of nuclear weapons. Higinbotham was named the first chairman and later its executive director. One of his first actions was the expansion of membership beyond just those who had worked on the Manhattan Project, which necessitated a name change, to the Federation of American Scientists. In a *New York Times Magazine* editorial, Higinbotham declared the reality of the bomb and outlined what the world had to do to avoid annihilation:

The longer we [scientists] lived with this problem the more alarmed we became. . . . The first thing we must understand about these forces is that against them there is no defense except world law. There will be no defense in the future, not until man is perfect. We must seek world control of atomic energy because it offers humanity its only measure of safety.[25]

Higinbotham served as the head of the FAS for two years before stepping down into an executive secretary role, but he continued as a tireless campaigner against nuclear proliferation until his death in 1994. It was only in the eighties, when Cold War tension between the United States and the USSR—and the threat of nuclear destruction—reached its zenith, that he felt his warnings were being taken seriously. "It's taken over thirty years, but the message is finally beginning to get through," he said.[26]

If he were alive today, Higinbotham would probably be disappointed to learn that he is better remembered for his contribution to video games than to nuclear non-proliferation. When the war ended, he went to work at Brookhaven as the head of the instrumentation team, which was responsible for building displays and measurement devices for the facility's research divisions. The lab, with its idyllic forest setting and mission to better control nuclear power, was the perfect home for the pacifist scientist. Over the next decade, Higinbotham and his team built all manner of electronic tools, from radiation detectors and a device that could measure a rat's heart rate to computer monitors that tracked the trajectories of missiles.

Residents of Long Island and New York, however, were uneasy about Brookhaven and the work that was going on there. Atomic research was a completely new science that had been shown to produce devastating results, and many people believed one slight mishap could obliterate the entire region in a mushroom cloud. In 1950 the facility tried to dispel those fears by holding annual public open houses to show that its research was in fact safe. The problem, however, was that much of the lab's work was top secret, so visitors were restricted to viewing photographs and inert equipment, which Higinbotham felt were boring.

Higinbotham was a fun-loving character, a chain-smoker who loved to liven up the lab's frequent parties with his frenetic accordion playing.[27] Many of the staffers at Brookhaven were out-of-towners who only socialized during weekend beach parties. Higinbotham went out of his way to keep their spirits up. He was also a self-confessed pinball junkie, which explains why he came up with Tennis for Two.

Although digital computers were starting to gain traction in the fifties, Higinbotham designed his game on an older analogue machine, which used on-off pulses to represent data. His box, about as big as a current-day microwave oven, was full of vacuum tubes and was programmed to simulate a game of tennis as seen from a side view, complete with varying ball trajectories and speeds. The computer was hooked up to an oscilloscope, a round screen about 5 inches in diameter that was normally used to display voltages. A green horizontal line represented the court, while a shorter vertical line in the middle was the net. Two small controller boxes were also hooked up to the computer; each had a dial that allowed the holder to direct the angle of the ball and a button that hit the ball when pressed. The entire design took Higinbotham three weeks to create. The result was effectively the first video game.

There were electronic games before Tennis for Two, but none of them was a proper video game. A 1947 game invented by American physicists Thomas T. Goldsmith Jr. and Estle Ray Mann simulated missile shooting, but required a plastic overlay to be placed on the screen because the computers of the time were incapable of drawing graphics. Similarly, in 1951 British firm Ferrant designed the Nimrod digital computer to play a game called "Nim," but the display consisted only of a bank of lights that blinked on and off. OXO, also known as Noughts and Crosses, designed in 1952 for the EDSAC military computer at the University of Cambridge, was perhaps the closest thing to a video game as it displayed graphics for tic-tac-toe on a cathode ray tube, but the X's and O's didn't move. Higinbotham's design was the first to feature moving graphics and incorporate what would become the three essentials of a video game: a computer, a graphical display and a controller apparatus.

When Tennis for Two made its debut at Brookhaven's visitor day on October 18, 1958, it was one of the first real, practical displays of what computers could do. While Art Linkletter had used the massive UNIVAC computer to match couples on his television show during the mid-fifties, Tennis for Two was the first opportunity for the general public to get their hands on one. Hordes of awestruck people lined up that Saturday afternoon to try out the game. Other lab departments looked on in envy as virtual tumbleweeds rolled by their deserted displays. The game was rolled out again the following year with some improvements, including a bigger screen and variable gravity settings to simulate what playing tennis on other planets might be like. After that, Tennis for Two was dismantled, its component parts put to other uses. Despite the game's popularity with visitors, Higinbotham never patented it or sought to commercialize it.

Robert Dvorak Jr., whose father built the game from Higinbotham's designs, doesn't believe the physicist knew what forces he had set in motion. "The whole idea was to show the public what a computer was, what it could do," he told me. "From the perspective of society, he had no idea what he was doing." Moreover, the cost of the equipment involved— about $20,000 by today's standards—precluded any thoughts of creating a commercial product. "The concept that this was something that could be within reach of Joe Consumer would never have occurred to anybody," Dvorak said.[28] Higinbotham himself later admitted that he never thought to patent Tennis for Two, and even if he had, the rights to the game would have belonged to the United States government. "We knew it was fun and saw some potential in it at the time, but it wasn't something the government was interested in," he recalled in 1983. "It's a good thing, too. Today all video game designers would have to license their games from the federal government."[29]

Dots, Blips and Blobs

The first landmark video game patent eventually went to Ralph Baer, a German-born Jewish inventor who escaped his native country just weeks before the Nazis launched the Kristallnacht pogrom in November 1938. Only sixteen years old when he arrived in New York with his parents and sister, Baer got a job repairing radios. When the war started, he was drafted and assigned to Military Intelligence, where he became an expert on small arms and won the army's Marksmen's Medal. While serving in England, he caught pneumonia and spent the latter part of the war in a military hospital. When he returned to the United States in 1946, the young veteran had difficulty getting into colleges because he hadn't finished high school in Germany. He finally caught a break with the American Television Institute of Technology in Chicago, which accepted him on the basis of strong entrance exams. Baer graduated with a bachelor of science in television engineering, a rare degree at the time. He then worked for a number of smaller firms, including satellite equipment manufacturer Loral Electronics, where he built his first television set.

Baer found his destiny in 1956 when he accepted a job at Sanders Associates, a defense contractor founded by former Raytheon employees in Nashua, New Hampshire, a tiny town an hour north of Boston. He was hired to manage the electronics design department and was soon promoted to oversee the entire equipment design division, a role that gave him a budget of several million dollars and more than 500 staffers. While his day job required him to work on airborne radar components and other defense electronics, Baer couldn't help but use some of his new-found resources to dabble in his real passion: television. While at Loral, he had become convinced that television sets could be used for more than just airing broadcasts. Indeed, he had managed to project test patterns onto the set he had built for the company in 1951, and found he could actually move the images around on the screen. That rudimentary concept—controlling images on a television screen—was the theoretical basis for video games.

At Sanders, Baer experimented with the idea further. In 1966 he created what he called the first-ever "television game." It was a simple demonstration that allowed two players to each control a dot of light on a blank screen with a handheld controller. The fun came from chasing each other's dot around—a concept that seems laughable by today's standards. Baer got semi-official backing from Sanders when he showed the game to his superior, who granted him $5,000 to further develop the idea. Two years and six designs later, Baer had his masterpiece: the Brown Box, a console that displayed a variety of rudimentary sports, maze and quiz games on a television screen. The system ran on batteries and the games were black and white. Most consisted of blobs of light moving around the screen; the controllers were big blocks with moving knobs. The console was hard-wired with all the games, and players switched between them by removing and inserting circuit cards, which connected different series of jumpers inside the machine. In homage to its creator's skill as a marksman, the Box also supported a plastic gun peripheral that shot at dots of light on the screen. As basic as it was, the Brown Box was the world's first video game console.

Selling it to Sanders's board of directors, however, was not easy. "Talk about long faces," Baer recalls. Only a few board members saw any future in the Brown Box, while the rest wanted to know how it would make money. "What the hell did I know about making money on commercial products?" Baer laughs. "I'd been in military electronics for ten years."

He finally convinced the board to support the console by spelling out the sheer potential market size—more than forty million households in the United States had television sets, and there were least that many in other countries. Surely some of those homes were interested in doing something more with their TVs.[30] Satisfied with his logic, the board gave Baer the go-ahead to shop around for a licensee to manufacture and market the console. Baer's first choice was the cable television industry, but he failed to find any takers. After showing the console to a host of television manufacturers, including RCA, Sylvania, GE and Motorola, he finally got a bite from Magnavox. The two sides signed a licensing deal and the Brown Box was rebranded as the Magnavox Odyssey. It hit stores in August 1972 with a $100 price tag, effectively launching the home video-game market, but it didn't meet with the success anyone was hoping for. Magnavox sold only a hundred thousand in its first year, mainly because of poor marketing; many consumers were turned off by the mistaken belief that the Odyssey worked only with Magnavox televisions.

The real money for Sanders and Magnavox didn't end up coming from console sales, but from patent infringement. In 1973 Baer got a patent for a "Television Gaming and Training Apparatus," which covered "the generation, display, manipulation and use of symbols or geometric figures upon the screen of the television receivers for the purpose of training simulation, for playing games, and for engaging in other activities by one or more participants."[31] Atari was the first violator when it launched its Pong home console in 1975. Pong was more successful than the Odyssey, primarily because of better marketing; the boxes were clearly marked "Works with any television set." Atari's success prompted a wave of Pong knock-offs from a host of companies, including pinball maker Bally Midway, which ultimately resulted in Sanders and Magnavox bringing a lawsuit to bear in 1976.

Sanders and Magnavox argued that Atari president Nolan Bushnell had ripped off their technology. And they had the smoking gun to prove it: Bushnell had attended a Brown Box demonstration in California in May 1972, three months before the Odyssey launched, as evidenced by his signature in the event's guest book. Bushnell settled with Sanders, becoming the company's first licensee outside of Magnavox. Baer, Sanders and Magnavox went on to sue virtually every company that attempted to get into the market, including Mattel, Coleco, Seeburg, Activision and Sega. They won or favorably settled every case after a slew of lengthy disputes. "They ran longer than any Broadway play ever did," Baer jokes.

The Brown Box inventor eventually went toe to toe with Higinbotham when Nintendo tried to invalidate Sanders's patent in the mid-eighties. Higinbotham was called as a witness by Nintendo to establish "prior art"—that Tennis for Two was in fact the first video game. The courts, however, again sided with Baer, who maintained that Tennis for Two was a simple oscilloscope-based ballistics demonstration and not a viable video-game system. The victory over Nintendo underscored once and for all that Sanders had first legal rights to video games, and that all companies dealing in the market going forward would have to pay the military contractor licensing fees until its patent expired in 1990.

Miniaturization Goes Massive

While many inventors turned from working on weapons to creating playthings for children, some got into an entirely different line of toys. One particularly key invention, the transistor, appeared right after the Second World War to lay the foundations of the technology industry, not to mention just about all the electronic toys we see around us today.

Leading the way was William Shockley, an American born in England and raised in Palo Alto, a small town south of San

Francisco. Shockley was by many accounts a terrible child, "ill-tempered, spoiled, almost uncontrollable, who made his doting parents' lives miserable."[32] Although he continued to be a difficult person into adulthood, Shockley proved to be a very smart and adept inventor. In the thirties, he got his degree in physics from the California Institute of Technology and his doctorate from MIT. Shortly before the war began he went to work for Bell Telephone Laboratories in New Jersey, where he was tasked with improving the vacuum tube, a device resembling a light bulb that could amplify, switch or otherwise modify electrons flowing through it. Vacuum tubes were the "brains" of many early electronic devices such as radios and huge, room-sized analogue computers, but they were clumsy processors at best. To relay the binary language of ones and zeros that electronics relied on, vacuum tubes had to be turned on and off—when they were on, they transmitted a one, when they were off, a zero. The process was slow and the tubes burned out frequently from the constant switching (or if an insect landed on a bulb, hence the term "bug," used today to mean "glitch").

Improving the tubes would have to wait, though. When the Second World War broke out, Shockley took an interest in radar development and toured the front lines to train pilots to use their new electronic targeting systems. Shockley was so respected for his mathematical prowess that, near the end of the war, he was asked to predict the casualties American forces could expect in a full-out invasion of Japan. His report ended up influencing one of the biggest decisions in human history: his calculation that a staggering 400,000 to 800,000 American soldiers would be killed in such an attack made the decision to drop the atomic bomb an easy one.

When the war ended, Shockley returned to Bell Labs and the vacuum tube. With his team, he experimented with different semiconducting materials to see which moved electricity most efficiently. The group eventually settled on germanium and gold, and in 1947 they unveiled the transistor, a chip made from these semiconductor materials which, after some improvements, worked far more efficiently and was much less fragile than the glass vacuum tube. No sooner was it invented, though, than a fight broke out over ownership. Bell Labs patented the device with John Bardeen and Walter Brattain, two of Shockley's team members, but left his name off the paperwork. There were also questions about whether the team had improperly referenced an earlier patent filed in Canada on a similar device, which was never built. A disgruntled Shockley, described by *Time* magazine as "a very competitive and sometimes infuriating man," wanted his proper due and set to work on an improved transistor, which he patented himself and unveiled in 1951.[33] Two years later, after being passed up for promotion at Bell Labs because of his difficult personality, he returned to his roots as a visiting professor at the California Institute of Technology. One of his friends there, Arnold Beckman, convinced Shockley to set up his own company as a division of his firm, Beckman Instruments. In 1955 Shockley Semiconductor opened its doors for business in Mountain View, only a few miles from Palo Alto, where Shockley had grown up.

Ironically, Shockley gave rise to a new epoch, not with something he did but with something he *didn't* do. After conducting some experiments with silicon as a semiconductor, he decided against using the material, much to the dismay of several of his researchers, who believed it to be superior to other substances.

Eight of his scientists, whom Shockley dubbed the "traitorous eight," split from his company over the decision and formed their own firm, Fairchild Semiconductor. In 1958 the new firm succeeded in building the first integrated circuit, which packed a number of transistors onto one miniaturized chip. (Texas Instruments, based in Dallas, coincidentally made the same breakthrough at about the same time.) This was the beginning of microelectronics and the official cornerstone of what would come to be known as Silicon Valley. Two of the "traitors," Robert Noyce and Gordon Moore, left Fairchild in 1968 to form their own company, Intel, in nearby Santa Clara. Today, of course, Intel is the dominant maker of microprocessors while Moore's famous 1965 prediction that the number of transistors that could be packed onto an integrated circuit roughly doubles every two years has become a "law" that still holds true.

The Silicon Valley companies ushered in a new way of doing technological research—and they added a business element to it. While many of their early efforts went to building computers, radar and other electronics for the military, the companies were civilian-run and profit-driven. They all had much bigger consumer markets in mind, which in the early seventies started attracting venture capitalists. Soon all the pieces for an electronics revolution were in place. With the nearby technologically minded Caltech and Stanford universities feeding them brainpower, venture capitalists pumping in funding, a healthy competitiveness and a burgeoning public appetite for electronic gizmos, the valley-area companies quickly discovered that silicon was indeed just as good as gold. Technology firms sprouted up by the thousands, not just to build silicon chips but also to handle the various spinoff businesses these created. Foreign electronics companies such as Germany's SAP and Japan's Hitachi moved in, and when the Internet started to take off, no one thought to locate themselves anywhere other than the valley.

Half a century after Shockley set up shop there, Silicon Valley is home to just about every major technology company in the world: Google (based in Shockley's old stomping grounds of Mountain View), Apple, Intel, AMD, Sun Microsystems, Adobe, Cisco, Hewlett-Packard, Oracle, Yahoo, Symantec, eBay and Facebook, just to name a few.

The transistor, meanwhile, is the father of the computer chips we know today. Many technologists consider it to be the greatest invention of the twentieth century. As one industry analyst puts it, "It has changed society. Look at transportation, computers, government, finance, manufacturing . . . it's affected them all. Look at the change in the productivity of the whole economy. It's probably doubled from what it would have been without transistors."[34]

It was and continues to be the lynchpin behind many electronics and has fueled the success and growth of many companies. Indeed, Japanese inventor Masaru Ibuka was so impressed with the transistor when he visited Bell Labs in the early fifties that he asked his country's government to help him pay the fee to license the technology. He brought the transistor back to Japan and used it to build a portable radio, which proved to be the first successful product for his small electronics firm. That company was Sony, and the rest, as they say, is history.

The scientific community recognized the impact the transistor would have soon after it was invented and honored

Shockley, Bardeen and Brattain with the Nobel Prize in physics in 1956. Shockley, never one to shy away from self-aggrandizement, said in his acceptance speech that the transistor was indeed the beginning of an entirely new way of thinking: "It seems highly probable that once the phenomena of surface states are thoroughly understood from a scientific point of view, many useful suggestions will arise as to how this knowledge may be employed to make better devices."[35]

Like many other brilliant but eccentric inventors, Shockley did much to ruin his legacy. While Slinky inventor Richard James converted to a religious cult and Tupperware designer Earl Tupper became a hermit, Shockley kicked his disgrace into gear when he started airing his views on eugenics. In the sixties he began to share his theories that intelligence was an inherited trait and that lower reproduction rates among smarter people were effectively making humanity stupider. Shockley took his theory further by saying that unskilled blacks had the highest reproduction rates in the United States and that over time, the black population would become even less intelligent than it already was.

To those ends, he donated his sperm to a bank devoted to preserving humanity's best genes and advocated paying people with IQs under a hundred to undergo voluntary sterilization. To say that Shockley's views were shocking would be an understatement. During the latter years of his life, he was largely written off by the media as a racist and often called a Nazi and a Hitlerite. He was hardly able to make any public appearances without demonstrators showing up. It all caught up to him in 1989 when he died alone, accompanied only by his wife, Emmy. His children found out about his death in the newspapers.

Still, it's hard to forget that William Shockley was indirectly responsible for many of the toys and gadgets we love today; and without him Silicon Valley might not exist.

War Becomes "Freaking Cool"

For decades, sociologists and psychologists have argued that violent video games and war-related toys such as G.I. Joe (made possible through the Barbie miniaturization pioneered by Mattel) are a deliberate form of psychological brainwashing, designed by the military-industrial complex to give boys a favorable attitude toward the armed forces. Others have suggested that a child's predilection toward war toys is hard-wired into the brain and develops from a sense of self-preservation where "instinctive animal play is practice for survival: the kitten's ball of yarn is tomorrow's mouse."[36] Playing with war toys, in other words, might prepare young children for the struggles they will face as they grow older. Still others have dismissed it simply as a "macho thing."

Me? As a kid I just really liked running around the forest and getting dirty, and I thought G.I. Joe was the coolest toy around, slightly ahead of Transformers. As I got older and my taste in toys became more sophisticated, I was attracted to video games. Like millions of other kids (and adults), I found them to be a great form of interactive entertainment.

Historically, the toy and games market has provided a much-needed creative, commercial and intellectual outlet for many military inventors and designers. Silly Putty and the Slinky, for example, were products of last-resort thinking, inventions with no practical use that were repurposed into playthings that turned out to be commercial sensations. In

Jack Ryan's case, Mattel provided the promise of fame and fortune as well as the canvas on which he could practise his technical creativity, attractions that building missiles at Raytheon just didn't offer. William Higinbotham, for his part, was driven to demonstrate that his work could not only destroy the world, but could also enlighten and perhaps even entertain it.

Recognition was also a main motivator for all of these inventors. Because of their military associations, they often had to conceal their work for security reasons. The Slinky and Tennis for Two were highly publicized, unlike the many other projects that Richard James and Higinbotham could not talk about. Scientists and engineers dream about toys and games because they are tangible examples of their work—products they can point to on a store shelf and show to their family or friends. As Ralph Baer says, "If you work at Sanders on a program for five years, it ends up as a box in an F7 fighter and nobody knows what you're doing because it's all classified. And even if you can show it off, it's usually a grey box. Toys and games are an attractive place to be."

The evolution of these toys has had a profound impact on how war is conducted. In its thirty-plus years of existence, the video-game market has mushroomed into an $18 billion industry; in 2007 it eclipsed the total revenue brought in by the movie business.[37] The military, for its part, has steadily stepped up its use of video games in training. In 2010 a new video-game unit of the U.S. Army, for example, spent $50 million to watch trends in the industry and identify technology that can be used to train soldiers.[38] Video games are also being used in recruitment, enticing potential soldiers with the allure of living out Xbox and Playstation war titles such as *Call of Duty* or *Ghost Recon* in real life. In 2009 the Air Force rolled out a sleek new website featuring interactive video games that allowed visitors to re-enact an actual A-10 Thunderbolt mission in Afghanistan, fly a Reaper unmanned aerial vehicle in Iraq or even refuel a plane in mid-air.

The technological development has come full circle— while many toys and games began as offshoots of military technology, they are now influencing and changing that same technology. When soldiers using Foster-Miller bomb-disposal robots complained that the dual-knob control system was too complicated to learn, the company redesigned it to use an Xbox controller. The same went for iRobot's PackBot, which now uses a Playstation controller. During a visit to iRobot's headquarters near Boston, I got to test-drive a PackBot, the company's own bomb disposer. Having grown up on video games, I had the machine mastered within minutes; I couldn't believe how easy it was to control.

The military knows its recruits today are video-game junkies, familiar not just with the technology but also with the violent themes, and it exploits this. "The Army will draw on a generation of mind-nimble (not necessarily literate), finger-quick youth and their years of experience as heroes and killers in violent, virtually real interactive videos," says one military journal.[39]

Video games are now providing playtime training to future troops, and those future troops don't even know it. This is dramatically changing new soldiers' views toward fighting wars, and not necessarily in a good way. The sociologists and psychologists who have argued that violent video games are desensitizing people to real-world violence may just be right, at least when it comes to the actual fighting of wars. As one young air force lieutenant described coordinating unmanned air strikes in Iraq, "It's like a video game, the ability to kill. It's like . . . freaking cool."[40]

FOOD FROM THE HEAVENS

Tasty Bite,
one of several
brands of
Indian food to
use packaging
designed
by space
agencies
and military
departments.

*The same intelligence is required to marshal an army in battle and to order a
good dinner. The first must be as formidable as possible, the second as pleasant
as possible, to the participants.*[1]
—ROMAN GENERAL AEMILIUS PAULUS

Does smelly, fermented cabbage sound tasty to you? Better get
used to it, because in a few years we could all be chowing down
on South Korea's national dish, kimchee.

The food, eaten by South Koreans with virtually every
meal, made its debut on the International Space Station in
2008, much to the delight of the country's first astronaut, Ko
San. "When you're working in space-like conditions and aren't
feeling too well, you miss Korean food," he said.[2] No kidding.
To South Koreans, kimchee is comfort food and a cultural
touchstone akin to pasta for Italians, apple pie for Americans or
Spam for Pacific Islanders. Many South Koreans attribute their
country's dramatic economic rise over the past few decades to the
invigorating powers of the cabbage dish. And while Westerners
say "cheese" when posing for photos, South Koreans smile and
shout, "kimcheeee!"[3]

Traditionally, kimchee was prepared in early winter, when
large clay pots filled with cabbage, seasonings and other vegetables
were buried underground to ferment. Today the process is more
advanced and kimchee can simply be bought at the grocery store,
then kept in a special refrigerator that regulates fermentation.

South Koreans eat it by the truckload, about 1.6 million tons a year, or more than 175 pounds per household.[4] Few non-Asians have even heard of kimchee, let alone tried it, because it doesn't travel well. The cabbage dish is full of microbes that help in the fermentation process, which means it has a short shelf life and is difficult to export. South Koreans abroad often find themselves missing their favorite food.

That looks to change with the work done by the Korea Aerospace Research Institute. Scientists there have found they can expand kimchee's shelf life by blasting it with radiation, which kills the bacteria after fermentation. The process also neutralizes some of the smell, which non-Koreans often find revolting. The result is "space kimchee," a safer, longer-lasting and less-pungent version of the earthly dish, ideal for consumption up in orbit. More important, the new creation will also have terrestrial uses, food scientists say. "During our research, we found a way to slow down the fermentation of kimchee for a month so that it can be shipped around the world at less cost," says Lee Ju-woon at the Korean Atomic Energy Research Institute, which began working on the food for the space agency in 2003. "This will help globalize kimchee."[5]

Eating Humble Pie

New and improved kimchee is the latest in a long line of food innovations to come out of various space programs. Exploring worlds beyond this one has meant overcoming a whole new set of technological obstacles, starting with launching humans out of the Earth's atmosphere. As we've ventured deeper, and as we've asked our spacefarers to leave their home planet for longer, the challenges of keeping them fed and healthy have become more complex. Space agencies have spent decades and millions of dollars in meeting these challenges, not just to keep astronauts and cosmonauts nourished, but also to provide them with a level of comfort so that they can concentrate on performing their scientific missions.

These investments, as is the hope with kimchee, have also paid dividends many times over back on Earth. In many instances, space agencies, particularly NASA, have transferred their technologies—whether new forms of packaging, processes or food chemistry—directly to consumer food companies, which in turn have used the advances to improve products. In some situations, food makers have developed their own intellectual property by working with the agencies, while in others whole new categories—such as camping food—have come straight out of space research. For much of the past fifty years, NASA and its kin have done as much to change the quality, cost and safety of food as the biggest terrestrial processors.

The whole notion of space travel traces its origins back to the military. While today we think of space exploration as a purely scientific endeavor and the ultimate example of international co-operation, that was definitely not the case in the early days of the Cold War.

At the end of the Second World War, the Soviet Union was at a big military disadvantage to the United States. Not only did the Americans have the atomic bomb, they had also recruited the best German scientists and engineers and a good number of the V-2 rockets the Germans had built during the war. Through the late forties, the U.S. Army quickly transformed the Nazi V-2 program, which was responsible for more than

2,500 deaths in Allied countries and a further 20,000 in German concentration camps, into its own space and missile programs.[6] As the fifties dawned, Americans were brimming with confidence—the country's clear technological superiority meant a safe and prosperous future lay ahead. A journey into space was merely a matter of when, not if.

On October 4, 1957, however, Americans had to eat a generous helping of humble pie (it was probably apple-flavored) when the Soviet Union, using its own German-captured technology and know-how, launched the first-ever man-made satellite, Sputnik I, into space. Despite its head start, the United States had been caught with its figurative pants down because of internal squabbles over funding and which branch of the military should have control over the space program. The surprise sent the American government and public into a tizzy.

In the fifties, launching rockets into space wasn't about who could venture farthest from Earth, but rather who could land nuclear weapons closest to their enemy. The Soviet Union had developed its own atomic bomb in 1949, but until Sputnik the American government wasn't terribly worried. Bomb-laden Soviet planes, while potentially deadly, could be detected and shot down well before they reached American territory. V-2 rockets, meanwhile, were only capable of making short flights, like from East Germany to the United Kingdom. No one knew yet how to fire a nuclear missile from one continent to another.

All of a sudden, the Soviet Union possessed that ability— and could wipe out the United States with the push of a button. The same, however, was not true in reverse. For the first time in history, the American people were faced with the very real possibility of annihilation by a technological superior. The doomsday clock neared midnight, and the Cold War shifted to a new level of urgency.

President Eisenhower ordered the formation of two agencies, the Advanced Research Projects Agency and the National Aeronautics and Space Administration, to ensure that the United States would never again be surprised on a technological level. We'll take a look at ARPA, which soon added "Defense" to its name to become DARPA, in chapter seven. NASA, meanwhile, was to be a civilian-run agency in charge of all aspects of space exploration and long-term aerospace defence research. The civilian veneer was designed to give the United States a sense of moral high ground over the secretive Soviet program, but there was little doubt that for much of the Cold War, the main motive behind any country's space program was to establish military superiority and hang the threat of nuclear annihilation over one's enemies, whether through rocket technology or espionage capability. That's why seven of the nine countries that have so far developed nuclear weapons (eight out of ten if you count Iran) have also launched rockets into space. It also explains why there is so much current concern over North Korea's attempts to shoot rockets into orbit.[7]

In 1961, before NASA and DARPA could get up to functional speed, the Soviet Union again beat the United States to the punch by making cosmonaut Yuri Gagarin the first man in space. NASA countered the following year by launching John Glenn who, during his five-hour-and-fifteen-minute flight, became the first human to eat in space while he was in orbit halfway between Australia and Hawaii. "I lifted the visor of my helmet and ate for the first time, squeezing some applesauce from a toothpaste-like tube into my mouth to see if weightlessness interfered with swallowing," Glenn wrote in his biography. "It didn't."[8] And so began the era of space food.

And the Bland Played On

Aside from the two T-38 Talon jets on display outside its main gate, there is little to distinguish the Johnson Space Center from any other American government research facility. The complex, consisting of a hundred or so buildings sprawled out over 2.5 square miles south of Houston, looks very much like a college campus. The rectangular low-rise buildings, linked by a grid of narrow, tree-lined roads, could conceivably house students learning socio-political theory or business administration. Instead, they're occupied by some of the sharpest brains around— scientists devoted to preparing humans for leaving Earth.

The only way for non-genius scientists to get past the guards at the front gate is to take a tour from Space Center Houston, the public visitors' building across the street, appropriately named Saturn Lane. After checking out garish, Disney-like displays of shuttle cockpits and moon rocks, visitors can ride a tram into NASA's facility to catch a glimpse of the agency's inner workings. Highlights of the tour include a visit to the sixties-era main control room, which was used during the Apollo program and now—with its push-button consoles, vacuum tubes and monochrome projector screens—looks like a kitschy set from the original *Star Trek* series. Visitors also get to see the cavernous training center, where astronauts prepare for missions by working inside full-scale replicas of the shuttle and space station modules.

Not on the tour, likely because it's nowhere near as sexy, is Building Seventeen: NASA's food lab. Here a dozen scientists dissect, formulate, test and create foods for consumption by astronauts on shuttle missions, the International Space Station and, perhaps soon, journeys to the moon and Mars. The main testing area looks like a cross between a cafeteria and a bachelor pad, with a large dining table set a few feet away from a kitchen counter. Various packaged foods are chaotically strewn across the room. A box-shaped contraption, like the automated detergent dispensers found in laundromats, sits at the end of the table. Dr. Michele Perchonok, NASA's food system manager, greeted me with a smile when I visited and revealed that the box was indeed an oven used on the space shuttle. My previous experiences with so-called space food amounted to eating the tasteless freeze-dried strawberries and ice-cream sandwiches sold at museum gift shops. I just couldn't believe that was what astronauts really ate, so I had to find out for myself.[9]

Perchonok served up a plate of beef brisket, accompanied by baked beans, cauliflower and cheese, mixed berries, cookies and, to wash it all down, a pineapple drink. I'd heard from talking to astronauts that the brisket was good, and it was indeed fantastic— the beef strips, flavored with tasty Texas barbecue sauce, were so tender that they seemed to melt in my mouth. The beans also had a nice smoky flavor while the cauliflower with cheese, despite looking like an unappetizing yellow blob, was savory too. I wasn't so thrilled with the berries, which were tart and lumpy, and the pineapple drink was the sort of run-of-the-mill powdered stuff you get at the grocery store. Still, the meal far exceeded my expectations. I was ready for total chemical blandness, the kind you get with pouched camping foods, but instead I got a meal that would pass muster in a decent restaurant. (NASA's brisket was actually on par with a plate I had later that night at the Goode Company Barbeque, a renowned Houston eatery.)[10] None of this was news to Perchonok, of course. "Mmm hmm," was all she said as I praised her cooking. Space food, which today is a mix of freeze-dried, dehydrated and irradiated products, has come a long way from applesauce in a tube.

NASA began developing its own food with the start of the Apollo program in 1961. With the goal of landing on the moon, Apollo missions would obviously be longer than the short Mercury and Gemini jaunts, so astronauts would need to eat. (Apollo 7, the first manned mission in the program, orbited the Earth for eleven days, while Apollo 15, the longest of the Apollo missions, clocked in at twelve and a half.) The problem was, no one really knew what to expect when it came to putting food in space. Microbes might mutate and become harmful, new kinds of bacteria might sprout up or the food might simply rot faster. There was also limited data on the long-term effects of zero gravity on astronauts' bodies. NASA played it safe and went with the most sterile and bland food it could find. In other words, army food.

The military had learned a valuable lesson during the Second World War: fighting an industrial-sized battle tends to work up an industrial-sized appetite. Fortunately, companies such as Hormel had stepped up to meet the military's food needs, even if it was with Spam. Though the post-war processing revolution resulted in longer-lasting and more portable foods, the military couldn't rely on private industry to put in the research and resources needed to meet its specialized requirements, which were likely to change with each new conflict. So Congress gave the Pentagon the green light to set up its own food science lab, and in 1952 the Quartermaster Research Facility opened for business in Natick, Massachusetts, a small town near Boston. The facility has added functions over the years and changed its name several times. Today it is known as the Army Soldier Systems Center, or more colloquially as the Natick Army Labs, and it supplies the military with food, clothing, portable shelters, parachutes and other support items.

At first, the lab developed standard canned rations that could be eaten in any battle scenario, but the tins ended up being too heavy and bogged troops down. One historian found that a special operations team could become "virtually immobile due to the weight of needed supplies . . . Mobility and stealth are decreased when loads become too heavy, and the soldier is too often worn down by midday."[11]

In the sixties, just as the space race was getting under way in earnest, Natick's focus shifted toward making lighter and more portable food packages, with a heavy reliance on dehydration and freeze-drying. Early versions of the Meal, Ready to Eat (MRE) became available to troops, to mixed reviews. The new rations contained a range of rehydratable foods, including beef hash, chili, spaghetti with meat sauce and chicken with rice. Soldiers complained about the taste, but were thankful for the reduced weight and simplicity, which validated the lab's approach. As the food-processing industry had learned in the fifties, making food that wouldn't spoil was easy—the hard part was getting it to taste good.

NASA scientists worked closely with their Natick counterparts to develop foods for the Apollo program, with the two labs refining freeze-drying and irradiation processes. Aside from weight and space considerations, the organizations discovered that they had much in common. NASA found that astronauts lost mass after spending time in space because there was no gravity resistance on their muscles. (Imagine not walking at all for a week; your leg muscles would become feeble from the lack of use.) The solution was regular exercise while in space. Today, astronauts on the International Space Station spend two hours a day on treadmills and other muscle-building machines to counter the effects of weightlessness. All that exercise requires extra calories, which makes astronauts similar to soldiers. Running around shooting at bad guys is tough work, so soldiers need about 3,600 calories, versus 2,000 for a regular (non G.I.) Joe.[12] That requirement contrasts significantly with the consumer industry, which has been under pressure for several decades to lower the caloric content of its foods.

Then, of course, there's the issue of longevity. "Our requirements fit more with the military's than the food industry's," Perchonok says. "We're both looking for longer shelf-life foods, shelf-stable food, which means they don't require refrigeration," since there are no refrigerators on the space shuttle or station. Dr. Patrick Dunne, Natick's senior advisor in nutritional biochemistry and advanced processing, says soldiers also need to be able to heft food around without refrigeration, so the military and NASA are both looking for foods with shelf lives of more than a year, compared with the industry's need of only a few months. "That makes our research environment a little unique compared to a commercial food producer," he says.

Although the two food labs have evolved in virtual lockstep, NASA has diverged from Natick in several ways. In zero gravity, astronauts generate fewer red blood cells, which absorb iron. Space foods must therefore be low in iron to prevent the mineral from being stored in other parts of the body, which can cause health problems. Weightlessness also causes bones to weaken, which means astronauts have to watch out for two imbalances: too little vitamin D and too much sodium. Even though they're considerably closer to the sun than we Earthlings, astronauts receive much less vitamin D because of all the heavy shielding on the spacecraft, so their diets must compensate. As for sodium, if you've ever looked at the nutritional information on a can of soup or a frozen entrée and the potential salt overdose they offer, it's easy to understand why NASA has avoided using commercially available products. With health consciousness growing among the public, food producers are starting to look to the space agency for help in decreasing sodium levels. "I have a feeling we're going to be working together in that," Perchonok says.

Poppin' Fresh Space Food

When Congress passed the 1958 National Aeronautics and Space Act, it insisted that NASA "provide for the widest practicable and appropriate dissemination of information concerning its activities and the results thereof" and "seek and encourage, to the maximum extent possible, the fullest commercial use of space."[13] In other words, the space agency was required to enrich American businesses by allowing them to profit from the technologies it invented. The specifics were outlined in the follow-up Technology Utilization Act of 1962. Taken together, the legislation served as President Eisenhower's not-so-secret weapon to make sure that the United States would never again be Sputnikked.

The amount of technology NASA has transferred to American industry has been, if you'll pardon the pun, astronomical. Never mind solar power and all the aerospace improvements, such as lighter-weight building materials, more efficient fuels and better sensor systems. There has also been a surfeit of medical gadgetry, including monitors for operating rooms that gauge patient oxygen, carbon dioxide and nitrogen concentrations, invented during the Gemini missions; bioreactors for developing new drugs and antibodies, from the space shuttle program; and micro-invasive arthroscopic surgery, made possible with technology from the Hubble telescope. The shuttle program has also improved roadways by introducing the idea of safety grooving, the cutting of tiny notches into concrete to increase traction; the process was first used on NASA runways. The Mars probe missions developed a new rubberized material five times stronger than steel for landers, which has since been used to add 10,000 miles to the tread life of commercial radial tires. On the consumer side, there's memory foam—used in everything from seats on amusement park rides to mattresses and pillows—Dustbusters, UV coating for sunglasses, friction-less swimsuits and even the Super Soaker squirt gun, invented at the Jet Propulsion Lab in California.[14] These examples are only the tip of the iceberg. Wernher von Braun, the former Nazi SS officer turned head of the early American space program, even made a contribution to consumer life by helping Walt Disney design his theme park.

The agency's technology transfer to the food industry has also been huge. One of the first fields singled out to benefit from space food research was health care. In the mid-seventies hospitals and nursing homes were suffering from a condition known as "tired food." Major medical institutions have to serve hundreds or thousands of meals a day, and there were often lapses between when the food was prepared in a central kitchen and when it was actually delivered. By the time the patient got his or her meal, it was often cold, tasted terrible and had lost many of its nutrients. NASA's solution was the "dish-oven," a hot plate–like contraption developed for the Apollo moon lander in partnership with Minnesota-based conglomerate 3M. The oven, which looked like an oversized soap dish, warmed food from beneath by zapping it with electricity. It was also highly energy efficient, as it needed to be for space missions, and used 60 percent less power than a regular oven. Moreover, it was small, lightweight and portable and could be set up in a patient's room, which decentralized food production by allowing meals to be warmed up on the spot.[15]

In 1991 3M refined the idea into the Food Service System 2, which stacked full meals on trays in carts that were then refrigerated. At mealtime, the carts were removed from their refrigeration units, wheeled to their respective floors and plugged in for heating.[16] NASA also piloted a project during the seventies called the Meal System for the Elderly in which it supplied freeze-dried food for homebound, handicapped and temporarily ill seniors. Oregon Freeze Dry, one of the agency's major suppliers, delivered its Mountain House meals such as spaghetti with meat sauce and Tuna à la Neptune, which were prepared by adding water, to 3.5 million people. The poor seniors—they turned out to be guinea pigs for what is now one of the most successful brands of camping food.[17]

It didn't take long for other major food companies to see the benefits of space technology. In 1972, with help from NASA, Chicago-based meat packer Armour turned a lunar lander strain gauge into the Tenderometer, a device that could predict the tenderness of meat. The company developed a ten-pronged fork that, when stuck into a side of meat, could measure the degree to which it resisted penetration. The device helped Armour market a successful premium line of beef known as TesTender.[18] Tip Top Poultry, meanwhile, used soundproof panels designed with NASA funding at one of its plants in Georgia, where high noise levels were degrading worker morale and safety. Conventional, plastic sound-absorbing panels weren't strong enough to stand up to the high-pressure water cleaning required by poultry plants, so the tougher fiber-reinforced polyester film developed for NASA to protect against vapors was a godsend.[19]

Other food makers were attracted to the actual fuel used to launch rockets into space. Liquid hydrogen, used by NASA because of its light weight and high energy output, turned out to be perfect for making margarine and keeping cooking oils fresh; it was also handy for pharmaceutical manufacturing and removing sulphur in gasoline production. In 1981 Pennsylvania-based Air Products and Chemicals, riding high off NASA contracts, opened a new plant in Sarnia, Ontario, to cater to this consumer market. "These applications would not exist today had it not been for our government experience," said the company chairman. "Our work on government contracts gave us the technological know-how for large-scale production of liquid hydrogen, enabling the cost reductions through economies of scale. That paved the way for expanded private-sector use."[20]

But NASA's biggest hit in the food-processing industry was HACCP, or the Hazard Analysis and Critical Control Point system. In 1959 the agency contracted Pillsbury, the giggling doughboy people, to create foods for the early Mercury and Gemini programs (and thus supply John Glenn with his applesauce).[21] Throughout the projects, the company discovered that its own food-testing methods were woefully inadequate compared to NASA's exacting needs. "By using standard methods of quality control there was absolutely no way we could be assured there wouldn't be a problem," a Pillsbury executive said. "This brought into serious question the then prevailing system of quality control in our plants. . . . If we had to do a great deal of destructive testing to come to a reasonable conclusion that the product was safe to eat, how much were we missing in the way of safety issues by principally testing only the end product and raw materials?"[22]

Pillsbury decided to completely overhaul its quality-control processes and reorient testing so that problems were detected before they happened, rather than after the fact. The company became the first American food processor to begin testing ingredients, the product, the conditions of processing, handling, storing, packaging, distribution and consumer use of directions to identify any possible problem areas. Pillsbury had its HACCP system in place for space food production by the time the Apollo program began and extended it to consumer plants shortly after the 1969 moon landing. The company then taught a course in HACCP to personnel at the Food and Drug Administration, leading to the publication of the Low Acid Canned Foods Regulations in the mid-seventies. The endorsements kept coming, with the National Academy of Sciences giving HACCP a thumbs-up in 1985, followed by the National Advisory Committee on Microbiological Criteria for Foods and the World Health Organization later in the eighties. In 1991 the U.S. Department of Agriculture's Food Safety and Inspection Service said HACCP was "the most intensive food inspection system in the world," while the company bragged that none of the 130 safety-related recalls between 1983 and 1991 were Pillsbury products.[23] The system was adopted as law in the United States during the nineties and in 1994, the International HACCP Alliance was formed to spread the standards worldwide. By the turn of the century, most major food growers, harvesters, transporters and processors in the developed world were working off some variation of Pillsbury's NASA-developed standard.

Judging a Food by Its Cover

NASA's technological contributions also spread directly to consumer food products. In the eighties, the agency discovered that a micro-algae it was testing as an oxygen source and waste disposal aid was actually a decent nutritional supplement. Scientists at Maryland-based Martek Biosciences found the algae produced docosahexaenoic acid (DHA) and arachidonic acid (ARA), rare fatty acids that play key roles in infant development and adult health. DHA is particularly hard to come by, as it is only found in breast milk. Martek came up with two nutritional supplements, life'sDHA and life'sARA, and marketed them to food companies. The supplements are now used by major food companies, including General Mills, Yoplait, Odwalla and Kellogg, and are found in products in sixty-five countries. An estimated 90 percent of all infant formulas in the United States use them, and about twenty-four million babies worldwide have consumed the algae.[24]

Space research has also helped speed up pizza and submarine sandwich preparation. In the nineties NASA contracted Dallas-based Enersyst Development Center to help design a compact and energy-efficient oven for the International Space Station.

The company came up with a new cooking technique called microwave-assisted air impingement, which blasts food directly with jets of hot air rather than warming the entire oven cavity. The technique cooks food faster—up to four times quicker than a conventional oven—and more consistently, so it retains more of its flavor and texture. Enersyst licensed the technology to food processors and commercial restaurants in the late nineties and by 2002 had more than a hundred thousand customers around the world, including the Domino's and Pizza Hut chains, where it cut cooking times from twenty-seven minutes to six.[25] The company also teamed up with home appliance maker Thermador in 1997 to offer the JetDirect, but this home oven never took off because its high price tag—more than $5,000— couldn't compete with the falling cost of microwave ovens. In 2004 Enersyst was acquired by Dallas-based TurboChef Technologies, which supplies Subway, Dunkin' Donuts and Starbucks with their high-speed ovens.

The consumer product that NASA and Natick scientists are most eager to discuss is one they jointly designed: the flexible "retort" pouch, which is finally starting to take off in grocery stores. The pouch, which is simply heated and cut open (or vice versa), is made from a plastic-aluminum blend and offers several advantages over canned goods. Like a can, it keeps out food's two biggest enemies, air and moisture, but because it's much thinner the food inside doesn't need to be cooked as long, which retains more natural flavors, textures and nutrients. This also means that fewer additives and chemicals need to be added to the food to keep it stable. And since shipping costs on food are calculated according to mass and volume, the pouch's lighter weight and more compactable form saves money for producers, which they can either pocket as profit or pass on to consumers through lower prices.

The metallized, foil-like material was originally developed by NASA to help bounce signals off communications satellites, but was then repurposed to insulate spacecraft from radiation and extreme temperatures. It's since been used in tents, rafts, blankets, medical bags and those reflective cardboard things you stick in your windshield in the summer to stop your parked car from turning into a solar oven.

Natick found the substance very handy, and after winning FDA approval for it in 1980, used it to create flexible pouches for MREs (Meals, Ready to Eat). NASA followed suit and now both labs use the pouches for most meals. North American food companies tried to sell products in pouches in fits and starts during the eighties and nineties, but none really took off, according to Natick's Patrick Dunne, because they adopted the same sort of drab packaging used by NASA and the military. The pouches did better in Europe and Asia because food producers there remembered that consumers actually care what packaging looks like—colorful and shiny sells, olive green with block letters does not.

North American producers have finally clued into that key tenet, and a flood of retort-pouched foods, from tuna and salmon to soups and rice dishes to fruits and vegetables—even Spam "singles"—has hit grocery stores over the past few years. With the American market for pouches growing at about 15 percent a year, it looks like they may yet replace cans.[26] "The graphics have really sold the product," Dunne says. "They did a nice marketing job." More important, Perchonok says, NASA and Natick made the pouches economical for food companies by performing all of the expensive research and development. "We've made that process a lot less expensive and got the packaging materials available at a price they can afford, so they are moving in that direction."

Nyet, Nyet, No Space Food Yet

What about the Russians? They've been launching into space for just as long as the Americans, so surely they must have come up with some pretty impressive food technology too, right? Like irradiated caviar or freeze-dried vodka?

Well, no. The Russian space program has taken a very different tack to NASA. The Soviets/Russians have generally used off-the-shelf canned goods, which has saved them millions on research and development of newfangled space foods. The extra weight incurred by the cans hasn't been a problem, because Soviet/Russian rockets have typically been bigger and more powerful than NASA's. The downside is that those bigger, more powerful rockets have required more fuel to launch, which costs more. On a pure cost-analysis basis, there's no telling whether Russia has ultimately come out ahead by not spending on food research.

As for the food itself, there have been few complaints from those who have actually had to eat it. Canadian astronaut Dave Williams, who went into orbit aboard Space Shuttle Columbia in 1998 and then up to the International Space Station for twelve days in 2007, was a big fan of canned Russian space foods such as caviar and borscht. "There are certainly downsides to using a can because once you take the food out you're left with it. Unless you have a trash compactor, the volume of your trash builds up correspondingly," he says. "But a number of the foods are spicy, which makes them quite palatable. I was impressed with the juices they had. Instead of being a crystal that gets water added to it, these were real fruit juices. It was remarkable to get to have those."[27]

Where Russia definitely lost out, though, is in the benefits of repurposing technology. Russian space historians say that during the Soviet era, the nation's space program was the least commercialized in the world. The Soviet Union didn't readily share its technology with industry, and industry didn't exactly beat a path to the space agency. One beaming example of this was the poor state of Soviet communications satellites, or "comsats," which had considerably shorter lifespans than their American counterparts—as little as two years, compared to the typical seven to ten for an American satellite. "For the USSR, the commercial imperatives were weaker and the competition non-existent: there were few incentives to build longer-lasting comsats," one historian explains.[28]

The situation wasn't helped by dramatic funding cuts after the fall of communism, with the space program losing about 80 percent of its budget between 1989 and 1999, making it one of the worst-funded in the world.[29] Russia's 2006 space budget—estimated at two billion euros—was, for example, a drop in the bucket compared to the 29 billion euros spent by the United States.[30] The country has also been late to the game in developing ties between its space program and private industry, only opening its Russian Technology Transfer Center in 2000, forty years after NASA did the same. "Such activities have never been organized in Russia before . . . the creation of RTTC will considerably facilitate export of technologies from the Russian Federation by making this process more organized," the center's director said in 2000. "RTTC specialists have studied very thoroughly the U.S. experience and legislation in the field of technology transfer. In fact, RTTC has been modeled after U.S. regional technology-transfer centers, particularly after the Office of Technology Transfer and Commercialization of the Johnson Space Center."[31]

Indeed, it took the arrival of Western fast-food companies to begin the process of modernizing Russia's food supply. McDonald's, which led the way, found the country's food system deplorable. "In Moscow, we had explored all sorts of meat plants and dairies and bakeries and had found that they weren't up to our standards," said McDonald's Canada president George Cohon, who spearheaded the company's move into Russia. "Nothing was easy. In the USSR of the late 1980s, the simplest things became logistical headaches. Could we get our bags from the Soviet Union? Could we get our napkins? Could we get our drinking straws? Even—could we get enough sand and gravel for construction? Could we get enough electric power?"[32]

McDonald's executives found that Russia and other parts of Europe were "light years" behind the United States in terms of food production. "The U.S. was twenty-five years ahead of many of our foreign markets in every aspect of food production— growing, production, distribution," one said.[33] (Energy, at least, didn't turn out to be a problem as McDonald's managed to get the Red Army to lay power cables for its new $40 million "McComplex" food processing plant.)[34]

Even after McDonald's laid down its roots in the early nineties and began overhauling Russia's food system—which included investing in farmers' equipment, irrigation, soil, transportation and distribution networks, not to mention the famous importation of potato seeds from the Netherlands—the country's agricultural system was still in a shambles. In the mid-nineties, officials estimated that about 70 percent of the nation's farms were on the verge of collapse. Analysts say change has been slow over the past decade and it'll be many years before the effects of technology—be it imported by foreign firms or developed internally through projects like the space program—will be felt.

The International Buffet Opens

Other countries are also lagging considerably behind the United States in their space food programs, although that's hardly their fault given the virtual American-Russian duopoly on manned space missions for most of the past fifty years. Space exploration only became a truly global endeavor in 1998, when work began on the construction of the International Space Station, which sixteen countries agreed to take part in. With long-duration missions now a possibility for much of the rest of the world, space agencies such as South Korea's are putting effort and resources into developing their own space food. Before long, they too will reap the sort of technology transfer benefits the United States has seen.

Japan, for example, only began researching space food technology in 2001. By 2007 the Japanese Aerospace Exploration Agency (JAXA) had twenty-eight items, including ramen noodles, green tea and teriyaki mackerel, approved for consumption aboard the ISS. Having Japanese foods aboard the station wasn't just a point of national pride, it also kept its crew comfortable and happy. "Our Japanese astronauts were concerned about American and Russian foods for the long-duration missions because most of these items contain meat and oils. The taste is too strong for us Japanese," says Shoichi Tachibana, chief of JAXA's health management team. "They were hoping we'd make some light-tasting foods. These light-tasting foods are good for the stomach."[35] Japanese scientists are now working on some more traditional national foods, including tofu and fermented beans, in the hopes of making them more shelf-stable and suitable for export, much like space kimchee.

India has perhaps the biggest head start, having launched air force pilot Rakesh Sharma into orbit through a joint Indo-Soviet mission in 1984. The Defence Food Research Laboratory, India's answer to Natick, used its expertise in developing military foods to supply Sharma with a variety of items, including curries, fruit juices, chapatis and chicken biryani. DFRL then offered its menu to NASA, which chose thirteen items for the Space Shuttle Challenger launch in 1986. Like NASA and Natick, India's military food lab transferred its technology, including its own version of the retort pouch, to local food producers, who have used it to export Indian products around the world. Western supermarkets today are full of Indian curries and rice dishes in retort pouches under brands such as Ashoka and Tasty Bite. Some of the packages even have labels that read, "Technology developed by Defence Food Research Laboratory, Ministry of Defence, Mysore, India."

With India now eyeing a manned moon mission, it is quickly ramping up food technology research, which is also becoming a national priority because of the country's explosive economic growth. With more and more Indians finding work every day, the country is mirroring what happened in the United States in the fifties, when ordinary people found less and less time to prepare fresh meals. "That necessitates the use of processed foods," says DFRL director Dr. Amrinder Singh Bawa. "They don't have much time to spend in the kitchen."[36]

China, which in 2003 became only the third country to launch its own manned space mission, is also devoting resources to space food technology. In 2007 the Scientific Research and Training Center for Chinese Astronauts made several items from its menu of sixty space-worthy foods, including chocolates and other desserts, available through grocery stores. Chinese experiments with fruit and vegetable seeds in zero gravity, meanwhile, have also returned some surprising results. When returned to Earth, the cultivated seeds have turned into giant fruits and vegetables with higher than normal vitamin content, a phenomenon that researchers say could solve the world's food problems. "Conventional agricultural development has taken us as far as we can go and demand for food from a growing population is endless," Chinese scientists say. "Space seeds offer the opportunity to grow fruit and vegetables bigger and faster."[37]

With food processing taking off in both countries, India and China are seeing a rapid influx of Western fast-food restaurants. Chains including McDonald's and KFC are seeing the same sort of growth in these new markets as they did in North America during the fifties and sixties. In 2004 there were 50 McDonald's restaurants in India and 600 in China; five years later there were 160 and 1,050, respectively. KFC had only 34 restaurants in India by 2009, but it dwarfs McDonald's in China with 1,900 restaurants, nearly doubling from 1,000 five years earlier.[38]

Pioneer Life in Space

If the current direction being taken by Natick and NASA is any indication, the future of food may ironically lie in less processing, not more. A technique Natick developed in the nineties using highly pressurized water and microwaves to cook and stabilize foods is now starting to catch on. The process involves sticking retort-pouched foods into a big drum, then adding water until the pressure inside builds to more than eighty thousand pounds per square inch. That kills off the micro-organisms in the food in about five minutes, whereas previous steam methods took over an hour. Again, shorter processing times means the food retains more of its natural taste, texture and nutrients, so fewer additives are needed afterward. Major food processors got their education on the technology through contracts with the military and are now starting to implement it. Texas-based Fresherized Foods was the first to have a hit with the process, using it to create its successful line of Wholly Guacamole dips. Even Spam maker Hormel is using the so-called "TrueTaste" technology in its preservative-free Natural Choice line of meats.

NASA, meanwhile, is now thinking about how to feed astronauts on a Mars mission. Such a trip will require food that has a shelf life of five years, significantly longer than most existing products (with the possible exception of the indestructible Spam). Then there's the problem of all that weight and waste: NASA's Perchonok estimates that sending six people on a thousand-day mission will require nearly 22,000 pounds of food, with more than 10 percent of that coming back as waste. "That may not be the most efficient way of doing things," she says. One likely solution is to send processing back in time and have crews grow the majority of their food in space. "It would be a gourmet kitchen, but with an 1800s look and feel because you wouldn't be able to go to the grocery store and get your grated carrots. You'd have to grate them by hand or by food processor." There you have it: the future of food processing is a cheese grater.

THE NAKED EYE GOES ELECTRONIC

Steven Sasson invented the first digital camera, a four-kilo monstrosity, at Kodak in 1975.

If the human body's obscene, complain to the manufacturer, not me.[1]

—LARRY FLYNT, *HUSTLER* PUBLISHER

In the early seventies, Lena Sjööblom never dreamed that there would be such a thing as the Internet, let alone that she would be the most important woman in its history. Yet a clock hanging in her living room bears the inscription "Miss Internet of the World," a title bestowed upon her by adoring fans.[2] While the world at large has never heard of Sjööblom, to the people who helped develop the Internet in its early, formative years—the imaging community—the former Swedish model is a legend.

Sjööblom's rise to geek fame began in 1969, when the vivacious eighteen-year-old native of Järna, a village southwest of Stockholm, set out on an adventure to the United States. Having just completed high school, she had no real plan save for visiting a cousin in Chicago. Once there, she found work as a live-in nanny. The job paid the bills and allowed her to absorb the American culture she had grown up watching on television and in the movies. She spoke English well and had no trouble making friends, including some photographers who were struck by the wholesome brunette, a rarity from a nation known for its beautiful blondes.

At the time, most photographers in Chicago fell into the orbit of the city's recently established publishing sensation, *Playboy*. While *Playboy* had begun humbly in Hugh Hefner's living room in 1953, by the early seventies it was one of the biggest magazines in the world, selling an average of 5.5 million copies per issue.[3] That circulation clout gave Hefner a commanding position in the local photography market.

Dwight Hooker, one of Hefner's mainstays, had already convinced Sjööblom to do some modeling work for catalogs and advertisements. With her laid-back, European attitude to nudity, she didn't need much prodding to agree to take some test shots for the magazine. For the adventurous Sjööblom, posing nude in *Playboy* was simply another all-American experience.

Sjööblom made her splash as the magazine's Playmate in the November 1972 issue in a pictorial entitled "Swedish Accent." The five-page photo spread depicted her in various states of undress, but also clothed and socializing with friends and family back in Sweden. In one non-nude photo, Sjööblom strolled past a poker-faced guard at Stockholm's Royal Palace with her friend Eva; in another she and her family celebrated Midsummer Eve by dancing around a maypole. (Given the pagan nature of the ritual, it probably would have been more appropriate to be naked in that one.) In the accompanying article, she heaped praise on her new homeland and criticized her native country; while the modeling work influenced her decision to stay in the United States, Sjööblom was more attracted to the freedom. "Though I miss my parents and brother very much and Sweden will always be my home, I couldn't move back unless the government changed," she said. "The country's becoming too socialistic for me."[4]

Many previous Playmates had basked in the media attention their magazine appearances drew, but Sjööblom fled the spotlight as soon as it hit her. Shortly after the issue was released she moved to Rochester, New York, to do some low-profile modeling work for Eastman Kodak catalogs. In 1977, with her short-lived *Playboy* fame faded, Sjööblom ended her American adventure and returned to a quieter life back in Sweden, where she found work as an administrator in the state liquor agency.

The seeds of her real fame, however, were sprouting in a small laboratory in Los Angeles, where the first significant research on digital image processing was taking place. Scientists at the University of Southern California had been working on transforming print photos into electronic formats since the early sixties, and by the seventies had established the school as a global leader in the field. In 1971 the Advanced Research Projects Agency, the Department of Defense section responsible for the development of new technology, acknowledged that status by giving the university a contract to digitize images so that they could be transmitted across its newly minted communications network, the ARPAnet. Two years earlier military engineers had successfully tested the network, the precursor to the Internet, and the Pentagon wanted to put as much data on it as possible, including photos. The university established a proper research facility, the Signal and Image Processing Institute, to tackle this challenge.[5] In a time before scanners and other readily available digital technology, SIPI's mission was daunting.

SIPI researchers built one of the world's first digital photo scanners at a cost of $250,000, a sizeable fortune at the time.[6] Newspapers and wire services had been using scanners since before the Second World War, but their devices were analogue and captured a much-degraded reflection of the photo. SIPI's digital scanner converted photos into the ones and zeros of binary code, which made it easier to manipulate them on computers and transmit them over communications networks. Researchers used the machine to scan photos of simple textures such as wood grains and leather, then aerial pictures taken from planes and spy satellites. From there, they moved on to more varied and colorful photos, including shots of trees, houses and jellybeans. The images were scanned and manipulated on computers through the application of algorithmic patterns, which bent, distorted, fragmented, disassembled, reassembled, blurred and sharpened them. By late 1972, however, the researchers were dying for a human face.

Alexander Sawchuk, then an assistant professor of electrical engineering at the institute, recalls that his team was desperate for a new test image because they "were tired of looking at all those other pictures."[7] One team member ran out to the nearest magazine store and picked up the latest *Playboy*, the fateful Lena issue. The magazine was chosen because it was one of the few publications that had full-color, high-quality glossy photos— Hugh Hefner insisted on using only the best photography and paper stock to avoid having his product considered a low-end skin rag—and its centerfold was ideal because it was the right size. Photos were wrapped around the scanner's cylindrical drum, which measured about five inches by five inches. Folded to hide the "naughty bits," the top third of the centerfold fit perfectly.

The picture featured a nude Sjööblom standing in front of a full-length mirror, looking back at the viewer over her right shoulder with long brown hair cascading down her back. She wore a floppy hat with a dangling blue feather, a pair of short black boots and stockings, and had a beckoning look in her eyes, punctuated by a mysterious Mona Lisa-like smile. Cropped to show only her head and shoulders, the picture was ideal for image research because it contained a range of color, had areas that were in and out of focus, and had alternating smooth and detailed sections. The skin tones were flat and simple while the feather on the hat was brimming with detail. It was the perfect candidate for the photo manipulation techniques Sawchuk and his crew had in mind.

SIPI first distributed the Lena image on tapes to researchers at other universities along with three other test photos: a pair of peppers, a fighter jet in flight and a colorful close-up of a mandrill's face. Sawchuk and his team, however, didn't tell anyone where they had found their sexy test subject. The picture was rescanned and finally transmitted over the ARPAnet in 1975.[8] At the time, the new field of image research was virtually all male and relatively youthful, so it was no surprise which picture got the most attention. Even a fighter jet couldn't compete with a photo of a beautiful, naked woman.

While the other three images were all but ignored, Sjööblom's picture quickly became the industry's de facto standard. As imaging publications noted in later years, it was impossible to work in the industry without being constantly exposed to the photo, which became known simply as "the Lena." Thumbing through industry journals in the seventies and eighties often turns up more than one Lena, sometimes dozens. "If the criterion is frequency of Lena, then the *IEEE Transactions on Image Processing* is by far the sexiest journal out there," read a 2001 newsletter from the Institute of Electrical and Electronics Engineers.[9]

The story behind the photo, like the best urban legends, was gradually forgotten. As the years wore on, scores of imaging scientists went about their daily algorithms without a clue as to where their sexy test subject came from. To them, she was "just Lena."[10] Like a modern-day Sleeping Beauty, however, the photo's past was awakened in July 1991, when it was published on the front cover of industry journal *Optical Engineering*. The periodical's editors didn't know the photo belonged to *Playboy*, and *Playboy*'s staff had no idea it was being used so widely by the research community. *Playboy*'s management sent a stern letter asking the journal's editors to seek authorization before using any of the magazine's images in the future, a request *Optical Engineering* was happy to oblige.[11]

As word of the photo's origins spread and political correctness crept into society and the imaging community, the controversy deepened and the objections multiplied, many stemming from the belief that *Playboy* exploited women. David Munson, editor of the *IEEE Transactions on Image Processing*, urged scientists in a 1996 editorial to use other pictures for testing purposes, both to broaden their studies and to placate the many researchers, both men and women, who had complained to him. Munson says his editorial had its intended effect—after 1996, use of Sjööblom's image dropped off. "People don't talk about it as much as they used to," he says.[12]

Nevertheless, by the mid-nineties the Lena had more than made its mark. In the seventies, it took several hours to transmit a photo over the ARPAnet. Using a good test image, researchers were able to refine compression algorithms to both shrink the size of files and speed up transmission technologies. In effect, they made the pipes faster and the data sent over them smaller.

Transmission times of electronic photos dropped exponentially as a result. By the late eighties SIPI's research had led standards bodies to agree on standardized formats for compressed images, including the Joint Photographic Experts Group (JPEG) and Graphics Interchange Format (GIF), as well as video compression with the Moving Pictures Experts Group (MPEG) standard.

Sawchuk is quite proud of that fact. When I visited USC, he showed me the original lab where the Lena photo was scanned. The space, on the third floor of one of the university's engineering buildings, its imaging equipment moved long ago, now sits largely unused and looks like any nondescript classroom. Still, Sawchuk beams with pride in showing it off. "We should have a plaque on the door that says, 'This is where the JPEG was born.'"

The JPEG became hugely important after the American government allowed phone companies such as MCI to connect their commercial email services to its ARPAnet in 1988, a move that effectively created the Internet. Electronic text and images were married shortly after, in 1993, with the launch of Mosaic, the first software program to combine the two elements on a single "page" that resided on the Internet. The pages became collectively known as the World Wide Web and individually as websites. Image compression work continued through the nineties and, combined with ever-faster network speeds, culminated in near-instantaneous transmission and loading of photos on the Internet and web by the early 2000s, with video not far behind. JPEGs, GIFs and MPEGs also became the de facto image and video standards on the Web.

Miss Internet of the World

On my visit to SIPI, Sandy Sawchuk told me a story that pretty much summed up the broad reach of "the Lena." In the mid-eighties he was visiting the state university of Novosibirsk, a Siberian city deep in the heart of communist Russia, when he was asked if he would like to see the school's imaging lab. "Would I? Of course I would," he told me. "So they showed me around and there of course was the Lena." I also showed the picture to Vint Cerf, the legendary computer scientist who made the first connection on the ARPAnet, and he instantly recognized it. Like many who work in his field, however, Cerf had no idea it was a *Playboy* image until I told him. Kevin Craig, *Playboy*'s digital lab manager, tells the other side: "I asked one of the IT guys here and I mentioned her name and he instantly knew who it was. I asked him, 'Is it more of an inside-IT geek thing?' and he said that's exactly what it is."[13]

The imaging community suitably honored Sjööblom, the Playmate who inadvertently influenced all those IT geeks, in 1997. Jeff Seideman, then-president of the Boston chapter of the Society for Imaging Science and Technology, tracked her down in Sweden and invited her to appear as the guest of honor at the group's fiftieth anniversary convention. By this point, Sjööblom had three children and several grandchildren and was going through a divorce. She was managing an office that employed disabled people to scan and archive corporate financial records, ironically using technology that traced its lineage back to SIPI. Seideman, a marketing specialist amid a sea of imaging scientists, convinced *Playboy* to capitalize on its former Playmate's fame. The magazine paid for her flight to Boston, and when she showed up at the convention, attendees were floored. The forty-six-year-old still had her figure, but her long auburn tresses were gone in favor of a short cut, now grey. Her eyes, however, still had that mysterious spark. "Many of them had spent their entire professional lives staring at her picture and had long since forgotten that she was a real person," Seideman says. "I introduced her to the main meeting and people were speechless. This creature that we dodged and burned and manipulated and sharpened and did obscene mathematical formulas to was a real human being."[14]

Sjööblom was just as surprised. She had no idea how famous her picture had become. She graciously met and spoke with convention delegates and signed autographs, including copies of her *Playboy* issue. Kodak, the firm that had produced the first electronic camera and for which she had modeled, set up a booth where convention attendees could have their picture taken with Sjööblom, creating a virtual Möbius strip for the digital age. Sjööblom returned home to Järna with fond memories. Aside from her family and a few friends, no one there knows of her fame. A self-confessed Luddite, she doesn't really understand the impact her photo had. "I'm not into the Internet," she told me over the phone from Sweden. She does, however, like the clock the imaging scientists gave her as a keepsake. "It's pretty cool."

Digital Shift

The same research that went into communications networks also paved the way for the development of many consumer electronic devices, notably digital cameras. Some of the earliest work on devices that could record images onto tapes rather than film began in the early seventies at Fairchild Semiconductor, the company started by

William Shockley's "traitorous eight." In 1973 Fairchild developed the charge-coupled device (CCD), a light sensor chip that would in later years serve as the brains of a digital camera in much the same way that a microprocessor powers a computer. Kodak licensed the chip, which converted what was seen by a camera lens into an electronic file, and quietly built the first digital camera—a hefty, nine-pound device—in 1975.

The camera took twenty-three seconds to record a single black-and-white photo onto a cassette tape, which was then played back on a television screen through a separate, additional console. Kodak patented the camera in 1977, but didn't put it into production because of its size and complexity.[15] Steven Sasson, the engineer who built it, later said the company had applied Moore's Law, which stated that electronics capabilities doubled roughly every eighteen months, to estimate when the digital camera would be compact and cheap enough to reach the general consumer market. Kodak's best guess was fifteen to twenty years, which proved to be fairly close to the mark: "but in reality, we had no idea," Sasson added.[16] The irony in Kodak's work on the electronic camera, which was kept secret for competitive reasons, was that Sjööblom—whose test image was doubtlessly used in the device's development—was modeling for the company, not that Sasson or his team could have done anything about it even if they knew. Security around the camera was so tight that no one outside of technical staff was allowed into his lab, nor was he permitted to take it outside. Besides, Sasson told me, Sjööblom's looks would have been squandered. "I had no idea that she was working there. I would have loved to have a model come and sit, but . . . they would have been wasted on the quality of the camera."[17]

Other camera and technology companies noticed Kodak's patent and got to work on their own electronic devices. By the eighties, some had made significant progress in improving picture-capture times, camera weights and image sizes. Sony made a major breakthrough in 1981 when it developed a two-inch-by-two-inch video floppy disc that allowed it to do away with tapes. Electronic cameras continued to improve throughout the decade, but by the nineties their photo quality was still nowhere near as good as film and they cost upward of $20,000. Still, there were some interested buyers, such as newspapers and the American military, both of whom deployed Kodak electronic cameras in the first Gulf War in 1991. The end of film really began in 1990, when Switzerland-based Logitech released the first true digital camera, which converted images to binary code, connected to a computer and stored images on a memory card.

The computer link finally allowed users to easily transfer photos from their cameras to their computers, where the images could be printed or sent over the Internet. Memory-chip capacities grew and image sizes shrank further through new compression techniques, which created a perfect inflection point in size and performance. Throughout the nineties, as the costs of charge-coupled devices plummeted and disc storage capacity rose steadily, digital cameras edged further toward the mainstream. Tokyo-based Nikon kicked the market into high gear in 1999 with the release of the D1, the first single lens reflex digital camera that was affordable to professional photographers. By the mid-2000s, the digital revolution was in full swing thanks to falling prices and continuing improvements in photo quality. In 2002 about 27.5 million digital cameras were sold, accounting for about 30 percent of the total still cameras shipped that year.[18] By 2007 digital cameras had all but killed off their film-based predecessors and essentially made up the entire market, with more than 122 million units sold.[19]

The key to the whole revolution was the ever-decreasing size of the photo files themselves, an advance that reached a high point with the JPEG and MPEG standards established in 1988. Like the Internet, all still cameras used JPEG as their standard file format by the mid-2000s, while video cameras relied on MPEG.

Eyes in the Sky

The U.S. military's return on its original investment in SIPI also bore fruit in the form of improved satellite surveillance. Before SIPI, the CIA had relied on spy photos delivered through its top-secret Corona project, an initiative launched in 1960 to spy on the Soviet Union, China and other regions of concern. Corona's satellites were equipped with high-altitude cameras that ejected spent film canisters, which parachuted to Earth only to be intercepted in mid-air by specially equipped aircraft. The canisters were designed to float in the ocean for a short time if they were missed in mid-air pickup, and then sink.[20] Amid the tensions of the Cold War, Corona was a necessary but hugely inefficient project—only about 70 percent of the 144 satellites launched during the project's twelve years returned usable imagery. There was also the constant risk that the parachuting film would literally fall into the wrong hands. The axe finally fell on Corona in May 1972, shortly before SIPI was founded, when a Soviet submarine was detected waiting below a mid-air retrieval zone.

The Landsat satellite program, which used sensors and cameras built by RCA and General Electric as well as an electronic transmission system that incorporated image compression started by SIPI, largely replaced Corona. Much of the Landsat project is still classified, so we can only guess at its military uses. Its commercial applications, however, have been widely publicized. Satellite imagery from the program has been purchased by agricultural, geological and forestry companies, and used by governments to predict and prevent natural disasters through the monitoring of weather patterns.

One of the program's first major commercial customers was McDonald's, which in its early days had scouted new locations by helicopter.[21] In the eighties, the fast-food chain converted to using satellite photos to predict urban sprawl. McDonald's later developed a software program called Quintillion that automated its site-selection process by combining satellite images with demographic data and sales projections. The software allowed the chain to spy on customers with the same equipment once used to fight the Cold War.[22]

Satellite photography was made broadly available in 1992, when Congress decided to sacrifice some of Landsat's secrecy in order to offset the project's cost. The Land Remote Sensing Policy Act declassified some of the data being produced and established that while "full commercialization of the Landsat program cannot be achieved within the foreseeable future . . . commercialization of land remote sensing should remain a long-term goal of United States policy."[23]

Landsat added a major new customer in the mid-2000s in the form of Internet search engine provider Google, itself no stranger to the military. Company founders Sergey Brin and Larry Page designed the algorithms that resulted in their ground-breaking search engine in the mid-nineties while they were students at Stanford University, right next door to Silicon Valley. The duo were the archetypal poor students, eating macaroni and cheese and begging for charity wherever they could get it. Among the handouts they received were computers from Stanford's Digital Library project, which was funded by the government's National Science Foundation, NASA and DARPA.

Google became the most successful company to emerge from the dot-com boom of the nineties by revolutionizing the Internet with its innovative search engine. After striking it rich by tying search results to online advertisements, the company moved to diversify its business in 2004 by acquiring Silicon Valley–based start-up Keyhole. The smaller company's main product was its EarthViewer 3D software, which used satellite imagery bought from Landsat and other commercial sources to create three-dimensional maps of the world. Keyhole was initially funded by Sony in 2001 and then backed by In-Q-Tel, a venture capital firm started in 1999 by the CIA to provide the intelligence agency with state-of-the-art spy technology.

The company was headed by John Hanke, who prior to receiving his MBA from Berkeley in 1996 worked in a nondescript "foreign affairs" capacity for the American government in Washington and Indonesia. When I asked Hanke what he did in foreign affairs during a visit to Google's headquarters, he wasn't exactly forthcoming. "Pretty much that. That's really the extent of what I've said publicly, so let's leave it at that," he told me with a grin.[24]

The company's main customers when it was acquired were the U.S. Army's Communications Electronic Command and the

Department of Defense. Keyhole, not coincidentally, was also the name of the satellites used in the Corona project. The company's software was renamed and relaunched in 2005 as Google Earth, a program that wowed Internet users with its lightning-fast rendering of satellite imagery. The pictures, benefiting from new super-fast computer processors, rapid Internet speeds, and of course, better image compression, loaded so quickly that they looked like full-motion video. Google Earth users also enjoyed the unprecedented novelty of zooming in on any location on the planet and viewing it in extraordinary detail.

Like virtually every Google Earth user, the first thing I did with the software was navel-gaze. I zoomed in on my home and was astonished by the level of detail, like the garbage cans out on the sidewalk. Obviously, I wasn't the only one who was amazed: "Since its debut on the Internet three years ago, Keyhole has had a high gee-whiz factor," wrote a technology reviewer for the *New York Times*. "When I first saw the site, I sat transfixed as it zoomed from an astronaut's-eye view of our planet down to a detailed shot of my house, with individual shrubs visible in the yard."[25]

Google expanded the software in 2007 to include Hubble Space Telescope photos of the moon, the constellations and Mars, and added the ocean floor in 2009. The company's Internet competitors, including Microsoft, Yahoo and MapQuest, were all forced to follow suit with their own three-dimensional mapping software, creating a boom in the commercial satellite photography market.

While the loosening of restrictions succeeded in creating a vibrant market for satellite photography, it also created headaches for governments and privacy advocates around the world. Google Earth's launch was immediately followed by media commentary on the software's potential negative effects, from complaints about invasion of privacy to concerns over national security. "Terrorists don't need to reconnoiter their target," said Lieutenant General Leonid Sazhin, an analyst for the Federal Security Service, the Russian security agency that succeeded the KGB. "Now an American company is working for them."[26] Google deepened the criticism in 2007 when it launched Street View, a feature that provides 360-degree panoramic ground-level views of city streets. Communities and privacy watchdogs around the world, including those in Canada and the United Kingdom, have raised concerns about Street View or passed outright bans of the software. Google relented somewhat in 2008 when it agreed to blur people's faces captured in Street View photos, but concerns about the company's further intrusion into daily life continue to swirl.

Home Invasion

Playboy's effect on imaging technology was accidental and indirect, but by the eighties, new technologies meant that the larger sex industry was exerting a much greater and more purposeful influence on the emerging home entertainment business. While cable television systems began rolling out in the United States in 1948, primarily to serve mountainous areas that couldn't get strong over-the-air reception, they only gained acceptance in the late sixties when competition emerged in the form of satellite TV providers. Cable penetration went from only 6.4 percent of American households in 1968 to 17.5 percent ten years later and 52.8 percent in 1988.[27] Rolling all that cable out, however, was costly for the providers, who needed quick revenue to justify the expense. Porn was just what the doctor ordered (or rather, what the companies' shareholders ordered).

At first, cable companies launched adult-oriented channels that customers could subscribe to for an extra fee. This set-up proved problematic, as morality groups argued that the content could be accessed too easily by children. A compromise was reached in the late eighties when cable companies switched to offering the majority of their X-rated channels exclusively through a pay-per-view ordering system, which effectively filtered out minors. Pay-per-view allowed cable providers to beam compressed, scrambled signals to subscribers; when the customer ordered a showing of a movie or event over the phone, the signal was unscrambled.

The technology first showed up in the late seventies and took off after a 1981 boxing match between "Sugar" Ray Leonard and Thomas "Hitman" Hearns. In 1985 a number of cable providers banded together to service the burgeoning market by launching several channels devoted exclusively to pay-per-view, including Viewer's Choice and Cable Video Store. Viewer's Choice II was launched shortly afterward to cater to a more mature audience with R-rated and soft-core pornographic films. Viewer's Choice II changed its name to Hot Choice in 1993 and, by 1996, was one of four adult channels, led by *Playboy* and Spice Networks, that controlled the business, collectively bringing in about one-third of the $600 million American pay-per-view market.[28]

The reason for their success was simple: for the first time, cable brought the product to the consumer, rather than the other way around. The porn buyer no longer had to sneak into a peep show in a shady part of town or sheepishly buy a magazine at the corner store, enduring the shopkeeper's disapproving looks.

Now the consumer could simply order pornographic movies and enjoy them in the privacy of his or her own home. The trend was repeated in the late nineties, when cable providers spent hundreds of millions investing in new fiber-optic networks that enabled two-way digital television communication. With new digital technology allowing cable subscribers to order movies and watch them whenever they wanted, rather than at pre-defined times, cable providers continued to look to porn to help pay for their investments. In 2000 porn brought in an estimated $500 million, more than 15 percent of all pay-per-view revenue.[29]

The same principle—bringing the product to the consumer—was responsible for the massive success of the home video market, which began with a war of technologies. Sony was first to market in 1975 with a home video cassette recorder (VCR), its Betamax player, followed two years later by the Video Home System (VHS) from Victor Company of Japan, or JVC. Netherlands-based Philips released a third format, the Video 2000, in Europe only. VCRs defied traditional rules of consumer electronics and flew off shelves despite the technology war, which carried with it the risk that buyers would be stuck with an obsolete product if one format lost the battle. The gadgets were also expensive; in 1979 VCRs sold for between $800 and $1,000, or about $2,300 to $2,900 in today's terms.[30] Still, more than 800,000 households had a unit before the seventies were through.[31]

Video players introduced several new experiences to the home. For the first time consumers could record television programs for repeated or later viewing, and buy and rent videotapes. At first, movie and television studios were apprehensive about licensing films to a format that could be either sold or rented, since this meant that their products could easily be pirated. In 1983 Universal Studios and Walt Disney Productions sued Sony and claimed the VCR encouraged violations of their copyrights. In what became known as the "Betamax case," the U.S. Supreme Court ruled in 1984 that making copies of television shows for time-shifted viewing was fair use and did not constitute copyright infringement, clearing the way for the industry's further development.

The studios later flip-flopped and came to see videotapes as a significant source of revenue, but in the early days they only grudgingly offered up their movies for transfer to videocassette. That left a void that was rapidly filled by porn. The June 1979 issue of television technology trade magazine *Videography* listed two porn titles, *Deep Throat* and *The Devil in Miss Jones*, among its top-ten-selling films (*MASH* led the list, followed by *The Sound of Music*). According to rental store owners, there was little doubt that porn was driving the business: "We're selling fifty times as many porno tapes as any of the other material," one New York store owner said. "We have all ages, all types . . . [most are men] but some women are buying the porno tapes too."[32]

Merrill Lynch estimated that half of all pre-recorded tape sales in the late seventies were porn, a percentage the genre held until mainstream movies finally caught up in the mid-eighties.[33] As with cable television, the secret to porn's success on the VCR lay in the fact that it brought the product to consumers, again saving them from visiting seedy or disreputable establishments. Most rental shops set up discreet sections where customers could browse porn titles. "There are some people who would like to frequent sex theaters, but for various reasons they don't. They're either ashamed to go in, they don't want to take their wives with them or whatever," said Joseph Steinman, president of Essex Distributing

Company, one of the largest porn producers in the seventies. "This way, they're able to see the X material in the privacy of their own home, and it doesn't seem so distasteful to them."[34]

VCRs also had a huge impact on the porn industry itself. Shooting straight-to-video was dramatically cheaper than using film. While the average feature-length film cost $100,000 to $300,000 to make, video brought that cost down tenfold. By the mid-eighties, the cost was again halved to around $15,000. The barriers to entering the pornographic film market were suddenly and dramatically lowered, leading to an explosion in production. In 1976 about one hundred new porn features were released; by 1996 that number had climbed to more than 8,000 a year.[35]

The sexual revolution may have begun in the sixties with the hippies and their mantra of free love, but it reached its climax— so to speak—in the late seventies and early eighties, when sex went fully mainstream. In the mid-nineties, New York mayor Rudolph Giuliani took credit for cleaning up Times Square by driving out all the drug dealers, prostitutes and pornography shops. The truth is, the job had already been half done by cable television, pay-per-view and VCRs, which had all allowed porn to go to the consumer, rather than the other way around. The shady, two-bit peep-shows that had proliferated in Times Square for decades were already feeling the pinch. By the time the nineties rolled around, porn was firmly ensconced in the home.

The VCR format war was finally settled when Sony threw in the towel on Betamax in 1988. JVC's rival VHS format offered longer running times and was cheaper to produce, and had gradually increased its market share to nearly 90 percent;

only about 21 million of the 170 million VCRs sold to that date were Betamax, while Philips had gained only a sliver of sales with its format.[36] In the aftermath, the war became one of the most studied cases in marketing and business courses, spawning a plethora of theories to explain why Sony had lost despite being first on the market with a strong product. Some analysts pointed to the influence of porn and the fact that, at first, Sony had refused to license its technology to producers on moral grounds. Porn producers also leaned toward VHS because it was cheaper. Others argued that Sony misjudged the market for VCRs, mistakenly assuming that the main use for the device would be recording television shows rather than renting and buying movies. As it turned out, consumers picked the latter use, for which VHS, with its longer running time, had the edge. Both theories are good, but regardless of which you favor the fact remains that in its early days, like so many new technologies, the VCR was supported by porn producers, who stepped into the breach created by hesitant mainstream studios.

Porn had a similar effect on video cameras, the first of which was released by RCA in 1977. By 1984 several of the largest American camera companies, including Kodak, RCA and General Electric, had partnered with Japanese manufacturers such as Hitachi and Matsushita (Panasonic) to release compact "camcorders" that sold for between $1,000 and $2,000.[37] The manufacturers marketed the gadgets as perfect for taping live events, but industry observers believed their main uses lay elsewhere. Assessing the emerging market for video cameras in 1978, *Forbes* magazine suggested that despite official marketing efforts, "it is an open secret that the biggest market is [visual sex]."[38] Indeed, most of the earliest camcorders had low-light capabilities, which didn't exactly lend themselves to filming sporting or family events. As Jonathan Coopersmith, a technology professor at Texas A&M University, puts it: "If you think about it, there are very few children's birthday parties which are really done with very low levels of light."[39]

Some entrepreneurs also cashed in on the popularity of using camcorders to film sex. Starting in the eighties, brothels offered customers "fantasy rooms" equipped with a video camera and VCR.[40] Fuelled by such uses, camcorders sold quickly. Sales for 1986 hit one million, nearly double the 517,000 sold in 1985. "The camcorder has quickly established itself as a potential billion-dollar business in only its second year of broad existence," said William Boss, vice-president of RCA's consumer electronics division.[41]

Prices continued to fall in the eighties, with the average camcorder costing less than $1,000 by 1986. That further spurred acceptance by mainstream buyers, who actually did use the gadgets to film children's birthday parties and baseball games. By the time the camcorder was ubiquitous, buyers with primarily sexual uses in mind made up just a small percentage of the total market, but porn's impact on the device's early development was undeniable.

Talk Dirty to Me

It wasn't all about the eyes, though. Along with cable television and digital photography, pornography spurred a good deal of development in phone technology, particularly in developing nations. Take Guyana, a small country at the northern tip of South America, for example. Today, the country's economy is dependent on agriculture and mining, with sugar, rice, bauxite and gold accounting for the majority of its exports, mainly to the United States and the United Kingdom. In the nineties, however, Guyana relied mainly on exports of an entirely different product to those same trading partners: phone sex.

It started in the United States in 1982 with a decision by the Federal Communications Commission (FCC) to end phone companies' monopolies on recorded messages. "Dial-a-porn" providers wasted no time in sprouting up, using new phone technology to offer customers recorded sex through 976 numbers for a fee. They also wasted no time in becoming immensely profitable: one phone-sex provider received 180 million calls in its first year and raked in $3.6 million.[42] In 1985–86, Washington-based C&P Telephone pressed the issue by saying it had not only earned $1.7 million through phone sex, it had also benefited the community at large by helping to keep basic calling rates low.[43] The story was similar in the United Kingdom. The industry began there in 1986; in 1987 it generated an estimated £80 million in revenue.[44]

Phone-sex services were easy for children to access, so the moral question soon cropped up. In 1985, New York–based Carlin Communications was the first to be charged by federal prosecutors for interstate transportation of obscene matter. Authorities tried to shut down the $2-billion-a-year industry in 1988 with a law that made it illegal to offer either obscene or indecent messages for commercial purposes. The Supreme Court, however, threw a wrench into the works a year later when it ruled that banning "indecent" phone calls violated the constitutional right to free speech. "Obscene" calls could be banned, but the discrepancy would have to be judged by local juries according to "community standards."[45] The FCC got involved in 1990 with new rules that forced providers to adopt age-verification technologies, including credit-card authorization, access codes and scrambling.

The court and regulatory actions, emulated in the United Kingdom, did little to shut down the operators—they simply moved them overseas. Phone-sex providers got around regulations by routing calls to phone banks in other countries, which then re-routed them to a third country where the call was actually answered. The migration began in the eighties, and by the end of the decade, international sex-related calls were pulling in nearly a billion dollars, or the equivalent of the entire domestic call market in Europe.[46] Phone-sex providers chose to route calls through small, poor nations because those countries needed the money and because the three-digit country codes of places such as Guyana, Moldova and the tiny Pacific island of Niue resembled North American area codes, so most callers didn't even realize they were dialing overseas. The system paid off handsomely for the developing countries; Guyana alone generated $130 million in 1995 from routed phone-sex calls, nearly 40 percent of its gross domestic product.[47]

Phone sex fell off considerably with the advent of the Internet, which catapulted the porn industry to another, much higher level. Family Safe Media, an online tracker of pornography, estimated that phone sex combined with cable, pay-per-view, hotel-in-room and cellphone porn to bring in about $2.6 billion in revenue in 2006, only slightly more than it was averaging on its own during its heyday a decade earlier.[48]

The decline, however, may be only temporary, as the rest of the world catches up in developing telephone systems. As China and India, the world's two most populated nations, add phones, phone sex is sure to follow. China is already facing the issue in its early stages, about twenty years after the United States and United Kingdom. In 2004 the Chinese government began a crackdown on phone-sex services, shutting down nearly a hundred operators in the second half of the year. "With the rapid development of the paid call service market in China, some lawbreakers make use of this form to spread obscene information and even conduct prostitution," said Wang Xudong, China's Minister of Information Industry. "This depraves social morals, and especially brings great harm to the country's young minds."[49]

At the close of the first decade of the twenty-first century, industry analysts expected Chinese phone-sex operators to follow their earlier American counterparts by moving offshore. The story there is only half written.

The spread of phone-sex systems in the developed world during the eighties and nineties had two far-reaching effects. First, phone sex allowed a number of developing countries with seemingly little going for them to attract foreign investment. Guyana, for example, benefited in 1991 when U.S.-based Atlantic Tele Network acquired an 80 percent stake in Guyana Telephone and Telegraph, the nation's biggest telecommunications company. Even though the phone sex wave largely came and went, Atlantic Tele Network has maintained its interest in GT&T and has since rolled out Internet and mobile phone services. While Guyana's Internet and mobile usage is low compared to the rest of the world, both are higher than they should be considering the country's GDP, which is among the lowest in the region. The initial investment in phone sex may have helped establish an enduring telecommunications legacy in the country.

In the developed world, phone-sex providers also pioneered a good deal of calling technology, including touch-tone-based credit card and access code verification. The next time you check your voice mail or order a pizza over the phone with your credit card, remember that the technology was first rolled out to give men a way to reach out and touch someone (or, rather, themselves).

In recent years, the term *phone sex* has also taken on a very different meaning. Porn producers are currently stampeding onto newer, more advanced cellphones that can access the Internet. Up until a few years ago, porn producers were largely at the mercy of cellphone providers, some of whom—particularly in sexually conservative North America—prevented their content from being accessed on moral grounds. In 2007 Telus, one of Canada's biggest cellphone providers, launched a service that allowed customers to download adult content onto their cellphones for a few dollars a pop, but shut the feature down a month later after it encountered outrage from the Catholic Church. That's why porn producers cheered the arrival in 2007 of Apple's iPhone, the first cellphone that successfully replicated the web-surfing experience of the desktop computer in a mobile format. While Apple refused to allow custom-made downloadable porn "apps" for the iPhone, the device's browser allowed users to view whatever websites they wanted, without interference from their cellphone provider. Porn producers quickly reformatted their websites for optimized viewing on the iPhone and the other smartphones it inspired. The market for mobile porn is consequently exploding. Analysts, who calculate mobile porn revenue separately from that found on the Web, expect the global market to hit $4.9 billion by 2013, more than double the $2.2 billion it earned in 2008.[50]

The reason for the big growth follows the logic of why VCRs and pay-per-view did so well in the realm of porn—cellphones not only allow the content to come to the consumer, they also allow it to come to them wherever they are and whenever they want. "You don't want somebody sitting next to you on the plane watching it, but sometimes if you're travelling and you're in the hotel room it's easier than getting out the laptop," says *Hustler* president Michael Klein.[51] Kim Kysar, brand manager for adult producer Pink Visual, sums it up simply: "It's the most private piece of technology you own."[52]

THE INTERNET: MILITARY MADE, PORN PERFECTED

American soldiers in Afghanistan pay tribute to Pink Visual with a bomb dedicated to the porn company.

The way you know if your technology is good and solid is if it's doing well in the porn world. [1]

—SUN MICROSYSTEMS SPOKESPERSON SUSAN STRUBLE

You may think a world without porn stars sounds like the distant past, but according to Tera Patrick, one of the biggest names in adult entertainment, it could also be the near future. That's because in the world of pornography, what the Internet giveth the Internet can also take away.

Born Linda Ann Hopkins to a Thai mother and British father in 1976, the raven-haired, olive-skinned beauty is a veritable legend in the business. This status is achieved through a combination of her exotic look, some entrepreneurship and a lot of hard work. (Believe it or not, being a porn star is not just about having copious amounts of sex and going to wild parties; there are also many long and dreary hours, tiring travel and repetitive publicity to contend with.) Hopkins's journey began early, when as a teenager she modeled for mainstream clothing catalogs. By the late nineties, she was posing nude for adult magazines such as *Hustler* and *Swank* and in 1999 she took the full dive into porn, changing her name to "Tera" to reflect her earth-consciousness and "Patrick" for her father's middle name.

Patrick is not the stereotypical sex-crazed bimbo. She earned a bachelor of science degree at Boise State University and was set to become a nurse, but found the potential riches of porn too hard to resist. Over the next decade, she appeared in nearly a hundred films, and in 2009 was inducted into the *Adult Video News* Hall of Fame, the industry's equivalent to the lifetime achievement Oscar.

After working for some of the biggest porn producers in the world, including Vivid and Digital Playground, Patrick struck out on her own with Teravision, a production company devoted to putting together what she calls a "beautiful, sexy video library." The Internet gave her all the necessary tools to get a business up and running, and in 2003 she launched ClubTera.com, a website where customers could buy online access to her photos and videos and order DVDs. ClubTera also allowed her to cross-link with other adult sites, promote personal appearances and interact with fans.

Talking to me from her office in Los Angeles, Patrick is upbeat and cheerful when discussing how she wants to be seen as a positive role model for other women in the industry. Her mood darkens, though, when she starts discussing her reasons for going solo. While the producers of her films, who were always male, capitalized on her popularity, the riches she had been promised never came. "The guys take advantage of the girls. The girls work for free or cheap and the guys make all the money. The producers make all the money," she snarls. "I was one of those girls who was working my ass off and making very little while the company took all my money . . . [Now] I'm one of the few women in the industry that does make money."[2]

This empowering equalization, what Thomas Friedman would call a "world flattener," extends far beyond porn into what is the most dynamic force in the world today.[3]

The Internet has given a voice to the disenfranchised and created opportunities where they never existed, from the very macro level right down to the micro. India, for one, has used the Internet to become the world's high-tech support center, transforming itself into one of the most important global economies after decades of poverty and hopelessness. Companies such as Google, eBay and Amazon, formed in garages and basements, have used the Internet to fundamentally change the way we live by literally bringing the world's information to our fingertips, or by allowing us to acquire all our earthly needs without ever leaving the house. At the lowest level, individuals—whether they are porn stars like Patrick or just simple teenagers—have been empowered to start their own businesses or blogs, or to connect, communicate and share with others through social-networking sites and tools such as Facebook and Twitter. The Internet has opened a new horizon of possibilities for people around the world, and its transformative abilities and effects are only beginning to be felt.

The Internet is also destroying the old ways of doing things. We're seeing the effects of this on an almost daily basis: if it's not a music label complaining about declining sales, it's Hollywood suing websites for disseminating copyrighted video or a newspaper going out of business. You may even be reading this text on your smartphone or tablet rather than on paper between two boards. Traditional industries are having to come to grips with the fact that in the new digital world, the old ways of doing business may no longer work.

This is also true in the porn industry, where anyone with a computer, a webcam and a high-speed Internet connection can now become a producer. The old production system, which revolved around the creation of big stars like Patrick, is under attack. On the consumer side, meanwhile, the porn industry is facing the same problems as mainstream media: its digitized content is increasingly being circulated for free. Having ridden the Internet porn wave to the top, Patrick understands that the system she came up in is being squeezed on both ends. "I don't know if there'll be another big porn star again because the market is too saturated and there's so many girls that no one really stands out," she says. "If you can go on the Internet and download free porn, why would you go to a girl's pay website, unless you were a big fan of hers?"

Bees in the Bonnet

It all started with President Eisenhower's other secret weapon against the Soviet Union, the Advanced Research Projects Agency. Like NASA, ARPA was formed in 1958 as a direct response to the Soviet launch of Sputnik, and also like NASA, it was a civilian-run agency, designed to oversee all advanced military research and to make sure the United States was never again beaten to the technological punch. Roy Johnson, a vice-president at General Electric, was appointed its first director and given an annual budget of $150 million. His ambitious vision included the creation of global surveillance satellites, space defense interceptor vehicles, strategic orbital weapons systems, stationary communications satellites, manned space stations and even a moon base.[4] The agency, which added a "D" for "Defense" to its name in 1972, was organized differently from a traditional research group in that it was small, decentralized and run by a handful of program managers scattered around the country. Tony Tether, who stepped down as agency director in 2009, was fond of saying, "Program managers do not work at DARPA, they are DARPA."[5] The managers, almost always leading scientists or engineers in their fields, have typically been recruited for four- to six-year periods and worked wherever they were based, traveling around the country as they were needed. Some observers have characterized the agency as "100 geniuses connected by a travel agent."[6]

ARPA's early priorities centered around ballistic missiles, developing technologies for space and detecting nuclear tests by other countries. By the late sixties, however, much of the work in those areas, including the Apollo program's Saturn rocket and the Corona satellite, had matured and was shuffled off to NASA and the relevant military and intelligence branches. ARPA's priorities then shifted to investing in what its directors like to call "the Far Side," or technologies that were still unknown or risky. Vint Cerf, whom many today revere as the "father of the Internet," was brought on board in 1976 as project manager for an ambitious new concept the agency was working on. The young engineer was an assistant professor at Stanford University, but he was enthused about building a worldwide telecommunications network, an idea that was only a dream at the time. He was a perfect match for the agency. "DARPA hires people who have bees in their bonnets and who want to make things happen," he says.[7]

Cerf is one of my favorite people to interview. He has a straightforward manner of speaking, and although he works for Google as its "chief Internet evangelist," he rarely sounds like he's preaching company spin.[8] Moreover, I love his wry sense of humor. During a speech to computer science students at the University of Waterloo in Canada in 2007, he flashed a slide of himself wearing a T-shirt that read, "IP on everything." To the nerds in attendance, it was comedic gold.

Nowadays, Cerf spends most of his time in Washington lobbying government officials to keep the Internet open and free. Google's digs in the capital are more serious-looking than its other locations, with fewer primary colors and less of the kindergarten feel the company's offices are known for. Cerf, the company's elder statesman, fits in well, with his trademark three-piece suit and impeccable manners. As usual, during our interview he was modest about his role in the creation of the Internet. The ARPAnet, its precursor, came about through the confluence of the ideas of three people, he explains. Paul Baran, an engineer working for the RAND Corporation military advisory group in the early sixties, was charged with "thinking the unthinkable"—designing a communications network that could withstand a nuclear attack. Baran came up with the concept of a "command-and-control" system based on the idea of hot-potato routing. Rather than just directing a phone call from point A to point B, his system called for redundancy, so the call would be distributed to a number of nodes in different locations— from point A to B and C and D and E, for example—and stored there. That way, if some of the nodes were disabled, say, by a nuclear attack, others would survive and communications could continue. Phone companies such as AT&T, seeing that such a network would be expensive to build, scoffed at the idea and told Baran he didn't know how communications worked.

The second prong came from J.C.R. Licklider, an MIT computer scientist who was appointed head of ARPA's

Information Processing Techniques Office in 1962. Licklider, or "Lick" to his friends, was interested in the psychology of using computers and in harnessing the power of many. Throughout the sixties, he developed his concept of an "Intergalactic Computer Network" in a series of memos that predicted just about everything the Internet has now become. If one mainframe computer was powerful, Lick argued, two computers working together were even more powerful, and the power only increased as more computing capability was added. He had trouble convincing MIT officials, though, who preferred to spend money on buying computers rather than connecting them.

The third prong was Welsh computer scientist Donald Davies, who came up with the term "packet switching" while working for the U.K.'s National Physical Laboratory. His ideas were virtually the same as Baran's, except they involved transmitting non-voice data in "packets" over a decentralized network. Davies built a small network in his lab but couldn't get funding to take the idea further either.

The prongs converged at a systems engineering convention in Tennessee in 1967 attended by all three parties. Notes were compared, common beliefs were discovered and before anyone knew it, Licklider's fellow MIT professor, Larry Roberts, was contracted by ARPA to head up development of a new packet-switching voice and data network. A plan was drawn up, the agency approved a budget of a few million dollars and the Massachusetts-based technology company Bolt, Beranek and Newman was hired to build it. UCLA and Stanford, where Cerf was working, were selected as the first two nodes for the network, and on November 21, 1969, the first permanent link in the ARPAnet was established. The first communication, however, didn't go very well—UCLA sent the word "login" to Stanford, but after receiving the letters "l" and the "o," the system crashed. Nevertheless, in a moment akin to the brief firing of the first nuclear reactor in Chicago in 1944, history had been made. For many of the computer scientists working on the project, establishing a network for defense purposes came second to creating new efficiencies in computing ability. "It had a command-and-control flavor to it," Cerf says, "but it had everything to do with sharing computing resources, because they couldn't keep buying new computers for everybody."

Selling Out

As more nodes were added to the ARPAnet and more data, like pictures of *Playboy* playmates, was sent across the network, a new problem emerged: the mainframe computers being used across the country in the early seventies were all built by different companies and ran on different software. IBM machines, for example, could only talk to IBM machines. It would be a full decade before personal computers started to take off and standard operating systems made by the likes of Apple and Microsoft started to gain traction. If there was any hope of expanding the ARPAnet into a truly universal network, which is what many of its creators hoped, it needed a common language to run on. That's where Cerf came in. After his appointment to DARPA in 1976, he came up with the Internet Protocol Suite, a common set of rules that determined how transmissions across the network worked. In order to get commercial computer manufacturers and network equipment makers to accept their protocol, Cerf and DARPA did the only thing they could do: they made it available for free. "We knew we couldn't possibly make this an international standard if we in any way constrained access to the technology, so even in the middle of the Cold War, DARPA flew under the radar," he says. "We kind of hoped the Russians would pick it up and it would occupy them for the next twenty years."

The network soon started paying dividends. DARPA found that using networked computers meant fewer machines were needed, so its mainframe costs shrank by 30 percent.[9] The rule of unintended consequences, so common when dealing with technological inventions, also reared its head. Scientists found the network was also good for sending text messages electronically. The "killer application," a term often used for the first practical application of a new technology, was discovered: email.

With most of the work done, Cerf left DARPA in 1982 for telecommunications company MCI, where he put together the first commercial email system. In 1983, with 113 nodes in operation, the American military split off from the ARPAnet into its own MILnet. The remaining ARPAnet nodes switched over to a new network constructed for academic use by the National Science Foundation, the NSFnet, which today sounds like something used to capture serial check bouncers. The final piece of the puzzle came in 1988, when regulators allowed Cerf's MCI email system and others like it to connect to the NSFnet. MCI made the link in the summer of 1989 and the commercial Internet was born.

The early Internet was mostly all text, but even at this early stage, porn was there first. By 1995 the text-based Usenet was dominated by sex and porn; chat rooms such as alt.sex made up four of the ten most popular bulletin boards, drawing an estimated 1.85 million readers.[10] Adult businesses helped develop uuencode, a tool that transformed lines of text code found on the Usenet into pictures, many of which carried ads for magazines and sex-chat phone lines.[11]

The Internet didn't really take off until it got a shiny new paint job in the form of the World Wide Web. Text was nice and all, but people really wanted to *see* stuff on their computers, especially when it came to porn. Tim Berners-Lee, a British-born MIT professor working at CERN in Geneva, got the ball rolling. In the late 1980s he put together some code called Hypertext Markup Language (HTML), which formatted text and electronic images onto a single page and allowed such pages to link to each other. On August 6, 1991, CERN uploaded its page, the very first thread in what Berners-Lee called the World Wide Web. The page contained links to information on how to create a browser that could view HTML code. Marc Andreessen and Eric Bina, two programmers working at the National Center for Supercomputing Applications at the University of Illinois, used the information to create their own web browser, Mosaic, which they made available to the public for free in 1993.

Mosaic, which eventually morphed into Netscape, was the first browser to seamlessly integrate text, graphics and links into an easy-to-use interface. Its release was also timed perfectly with the first generation of university students graduating and going off into the real world. As Cerf puts it, "People left university and they'd say, 'Where's my Internet? Where's my connection?'" The stage was set for the Internet and the web to take off, and the porn people were waiting. Berners-Lee knew it. "We have to recognize that every powerful tool can be used for good or evil. Legend has it that every new technology is first used for something related to sex or pornography," he said years after unleashing his invention. "That seems to be the way of humankind."[12]

A Playground for Porn

With the launch of its website in August 1994, Playboy Enterprises was one of the first big businesses to embrace the web. Company executives say *Playboy* was the first national magazine with a website.[13] The reasoning was pretty simple— porn is very much a visual medium, and the web removed the biggest obstacle to selling pornography and sexual services: the shame of being discovered.[14] Not surprisingly, people flocked to sites that displayed nude pictures. On its launch day in March 1995, *Penthouse*'s website received 802,000 visits. By 1997 *Playboy* was racking up five million visits a day, making it one of the most popular sites on the web.[15]

This instant success, however, meant that the big adult magazines were also the first victims of the problem that has plagued the Internet since its inception: copyright infringement. Even before the web, enterprising computer programmers were digitizing *Playboy* and *Penthouse* photos for transmission over the Usenet (and image researchers were downloading the Lena picture). The web made it even easier because anyone could copy pictures from *Playboy*'s or *Penthouse*'s site and start their own clone website. The solution for the magazines was to innovate, both technologically and legally. In 1997 *Playboy* unveiled technology that inserted a digital watermark into its images. The watermarks were invisible to the naked eye, but could be detected by a "spider" tool that crawled the web looking for them. "The worm is very appealing," a magazine executive said. "We try to see who's doing what, where and when to our stuff."[16] Every major content producer on the web now uses some variation of this sort of technology. On the legal front, in 1998—well before the big music and film industries went after copyright infringers with lawsuits—*Playboy* won a $3.7 million judgment against a California operation that was selling the magazine's images on the Usenet for $5 each.[17] All of a sudden, *Playboy* was in the unusual position of suing someone else for distributing pornographic material—its own!

Photographs would only do for so long, though, and soon porn entrepreneurs turned to developing online video. A full decade before YouTube, Pythonvideo.com was streaming live video from a number of sex theaters in Amsterdam's red-light district. Launched in 1995, the endeavor was limited by the early video compression standards and slow Internet speeds of the time, but the small, choppy streams proved immensely successful. Originally intended as a promotional vehicle to encourage tourists to visit the theatres, by 2001 Python was streaming live video content to 3,000 websites.

Other players, such as Virtual Dreams, took Python's idea a step further and introduced two-way video-conferencing strip shows. Visitors to the site typed in their requests to the man or woman at the other end, who then acted them out. Presciently, Virtual Dreams' owners predicted the technology would benefit "medical, educational and a whole host of commercial and industrial transactions."[18] Of course, video chat is now a standard tool offered by the likes of Apple, Google, Skype, Microsoft and Yahoo. Burger King even copied the idea years later with its hilarious "subservient chicken" website, where a man dressed in a chicken suit took requests from visitors.

Into the new millennium, several porn producers, including Los Angeles–based heavyweight Wicked Pictures, pushed the adoption of better compression standards, particularly the H.264 codec (also known as MPEG-4)—now widely used in online video—to take advantage of the higher Internet speeds being made available to consumers. "We don't sell millions of copies of a title on hard goods like a mainstream studio does, so we choose to support new avenues to deliver our products," Wicked founder Steve Orenstein says. "We embraced H.264 early on as our primary codec for online delivery for both archived and live high-definition streaming."[19]

The porn industry's video innovations did not go unnoticed by mainstream businesses, which quietly adopted them for their own purposes. One senior executive who would prefer to remain anonymous told me of how the multinational bank he worked for co-opted porn technology to help advise clients on their investments.[20] In 2003 the bank was looking for a way to distribute a video multicast of each morning's investment tips, but no one was offering such a service commercially and the bank's own IT research team was stuck. "Then we said, 'Hold on, having a talking image distributed by the Internet is something the porn people have been doing for a long time,'" the executive says.

The IT team came up with the idea of having an employee sign up to various porn sites to snoop around, but the bank's human resources department was, not surprisingly, "dead against it." A junior member was sent home armed with a credit card, instructions to expense all his research as meals and a mandate to figure out how the porn companies were broadcasting their videos. He came back a few weeks later with a fully functional video system. "It was a direct analogue. It was completely the same approach and the same technology," the executive says. "The higher-ups didn't care, they were completely in on the game and used to just laugh. They were completely conflicted."

Mainstream businesses also latched onto the security methods developed by porn companies, which by the nature of their business had huge targets painted on them. "The mixture of how 'adult' is seen as less accepted in society and the idea that money flows like water through this industry makes it an ideal target for hackers," says Paul Benoit, chief operating officer of the company that runs Twistys.com, a popular video download site. "Adult companies are less likely to work with the FBI or RCMP in dealing with hacker attempts since the nature of their business isn't as accepted as banking or book selling."[21]

Aside from having to deal with the same issues as mainstream websites, such as denial-of-service attacks, worms and viruses, adult sites also have to prevent hackers from stealing their content, which is usually hidden behind a paid membership wall. "This includes making sure our members area is locked down to authorized and paying customers only, but we also have to have defenses in place to prevent password sharing, password brute forcing, proxy server abuse and such," says Jack Dowland, systems administration director for a number of adult sites, including Pink Visual. "Everyone wants something for free. And not just free stuff for free. They want stuff that you would normally have to pay for for free. Add in the technical challenge, and you have a playground ripe for would-be crackers."[22]

Porn sites have thus been a target for every horny fifteen-year-old computer geek with too much time on his hands. As with *Playboy* and its watermark software, porn sites have had to make big investments early to protect their content and customers' private information. For many, security has become almost as important as producing the sexual content. As such, while having "porn star" on your résumé may not get you acting gigs opposite Clint Eastwood or Meryl Streep, having "porn webmaster" can result in a warm welcome from mainstream companies looking for innovative IT employees. That's not to say that many porn webmasters want to jump the fence—they tend to get paid well and enjoy more freedom at what are almost always smaller, more flexible companies. Working in the porn industry, Twistys's Benoit says, means doing "whatever our imaginations can generate."

While many web developments came from companies that simply made erotic films, the darker and creepier side of human sexual desire has also, unfortunately, inspired big advances. The arrival of Java, a programming language that made possible many multimedia applications, is one such example. Patrick Naughton was part of the team at Sun Microsystems that designed Java in the mid-nineties. In 1999 Naughton was nabbed by an FBI sting at the Santa Monica pier in Los Angeles, where he arrived for what he believed would be a sexual rendezvous with a thirteen-year-old girl he had met online. Naughton, who at the time was overseeing Disney's Internet content, was convicted of traveling across state lines to have sex with a minor, but avoided jail time by brokering a deal to help the FBI capture pedophiles online.[23]

Java, meanwhile, proved to be a great tool for web designers to create all sorts of in-browser applications, from games to interactive weather maps to real-time chat functions. It also allowed designers to create new functionality without having to worry about how to get past the firewalls used by many corporations, which tend to block employees from downloading add-on software for their computers. This proved to be a huge boon to the porn industry, which sees about 70 percent of its traffic happen during the nine-to-five work day.[24] Observers within the IT industry, meanwhile, have for years speculated about why Naughton was so interested in creating communications tools that could bypass corporate firewalls.

Show Me the Honey

The porn business is a business, after all, and none of these innovations would have happened if there wasn't a whole pile of money to be made. To cash in, adult companies were quick to develop a variety of payment systems, some good, others bad. Aside from adapting the automated credit card payment systems they had developed for phone sex, porn producers also invested in things like e-gold and OmniPay, digital currency transfer systems that allowed people to pay for goods online without a credit card. These systems have fallen into disrepute in recent years because they've become the primary payment vehicles of online casinos, which in North America are largely banned from accepting traditional financial transactions. (They did, however, inspire well-regarded systems such as eBay-owned PayPal.) One of the shadiest systems required customers to download and install special software. In one of the most insidious scams the Federal Trade Commission has ever seen, the software then made its own connection to an Internet provider in Moldova where, unbeknownst to the customer, it racked up huge long-distance phone charges. Customers only found out when their phone bill arrived.[25]

On the plus side, porn companies also pioneered co-operative affiliate systems where one website advertised on another. If a visitor followed a link on Site X to Site Y and ended up joining Site Y, Site Y then gave Site X a payment for the referral. Not only did mainstream businesses such as Amazon and eBay co-opt this innovation, it also formed the basis for Google's entire context-based advertising system. When you enter a query on Google, a number of text ads pop up on the right hand of your search; if you click on one of those ads, the company behind it sends Google a payment for the referral. In ten short years, Google has used this ad system to become the massively profitable behemoth it is today.

Regardless of the payment systems used, putting porn on the Internet was—and is—akin to printing money. The traffic and revenue numbers have been simply huge, even if they have been hard to accurately measure, since most producers are not large, publicly traded companies required to report earnings. Still, analyst estimates have painted an astounding picture. In the early days of the web, sexually oriented products accounted for an estimated 10 to 30 percent of the entire online retail market.[26] By the turn of the millennium, while mainstream content providers such as *The Wall Street Journal* charged online subscribers $59 a year, adult sites such as Danni's Hard Drive were able to charge $25 *per month*, which explained how porn accounted for more than half of the estimated $2 billion spent on online content in 1999.[27] The most profitable non-sex category of websites, online gambling, took in only $150 million in 2000, a paltry amount compared to the $1.7 billion raked in by adult sites.[28] Mainstream businesses also quietly benefited from the heavy traffic being generated by porn. Of the 81 million people who accessed popular search sites

Yahoo and MSN in March 2001, more than 30 million made their way to an adult site.[29] At the same time, about 40 percent of Germany's and Italy's entire web traffic was aimed at porn sites, with similar numbers found across Europe.[30] That traffic helped the portals sell advertising space on their sites.

Internet service providers also reaped the benefits as customers converted their slow dial-up connections to more expensive high-speed broadband. In the United States, about 20 percent of AT&T's high-speed customers paid to watch porn online. In Europe, where high-speed subscriptions grew 136 percent in 2002, music file-sharing services and adult content were identified as the two leading reasons for why people switched to broadband.[31] "Adult content is the obvious subscriber service to go for because there is already a proven business model," one analyst said at the time.[32] The revenue growth in online porn continued through the mid-2000s, particularly in the United States, where producers raked in close to $3 billion of the total $5 billion global porn pie in 2006.[33] In 2009 an estimated 25 percent of all search requests were for adult content while a third of all websites were pornographic, garnering as many as 68 million hits a day, or 28,000 surfers a second watching porn of some kind.[34] Today, $89 is spent on porn *every second*.[35] Even taken with a grain of salt, the numbers are astounding.

Too Much of a Good Thing

But this money-printing business is now under attack. A recent estimate by *E-commerce Journal* pegged the online porn market at $2 billion, or about the level it was at in the early part of the 2000s.[36] DVD sales and rentals, meanwhile, are down by about 15 percent. For the first time in its history, the supposedly recession-proof adult entertainment business is contracting in terms of revenue. *Adult Video News*, the *Wall Street Journal* of the industry, attributes the decline to the same two factors that have Tera Patrick worried: too much product and too much piracy. "The laws of supply and demand have been turned upside down. We're on par to put out fifteen thousand new releases this year, which is just insane," said *AVN* founder Paul Fishbein in 2008. "Secondly, there's a battle with pirated or free material on the Internet. Much like the music industry, adult producers are trying to figure out how to stem free or pirated content."[37]

Ironically, the product glut is the result of Internet innovation. With the ability to create and distribute videos more cheaply and easily than ever before, everyone with an Internet connection can now be considered a competitor to the likes of Vivid and *Hustler*. Piracy, meanwhile, is coming in two forms: file-sharing and free websites. The mainstream music and movie industries have felt the damage of websites such as Pirate Bay and Isohunt, which contain directories of "torrents" that allow users to swap files for free. The potential hurt could be much worse for the porn business since, as Wicked's Orenstein said, it has no other revenue stream—such as live shows or box-office earnings—to fall back on. Even worse are a rash of porn-flavored YouTube clones, including YouPorn, RedTube, XTube, and Tube8, which allow users to upload sex videos. The sites are drawing considerably more traffic than the big porn companies' own online operations, and they're rife with copyrighted material. That has the producers hopping mad. "This needs to be treated like a bank, like someone is going in and robbing a vault. As far as I'm concerned, for the music business, for Hollywood and for our industry it has to be treated the exact same way," says Digital Playground president Samantha Lewis. "We'll all be out of business at this rate."[38]

The producers' retaliation has come in several forms, some more successful than others. The bigger companies have done remarkably well in keeping their content off torrent and tube sites, for example. While Warner Bros. declared victory in 2008 by announcing it had kept its blockbuster Batman movie *The Dark Knight* off torrent sites for two whole days after its theatrical release, whereupon it became easily available, the big porn companies are taking comfort in the fact that many of their films are file-sharing rarities. Lewis claims that this is the result of a persistence not practised by Hollywood. Digital Playground, for one, has two full-time employees monitoring torrent and tube sites who take immediate action against any and all infringements. "All we do is shoot off an email and it's down in thirty seconds. They really don't want a lawsuit," Lewis says. "We'll get something down and then they'll mix it around, maybe change the spelling of the name and put it right back up in fifteen minutes. We'll be on them again. If it happens again, then they get the letter [from lawyers]. You just have to be persistent and they know we're serious."

Some feel the ultimate solution to stopping piracy lies with what brought the industry to the dance: innovation. Stoya, winner of several "new starlet" awards in 2009 and one of porn's current "it" girls, says lawsuits and technological restrictions like copy protection haven't worked for the music and movie businesses, and they won't work for adult entertainment. Maybe, then, it's time to get creative. "It's not so much that they're fighting a losing battle, but they're fighting a different battle. The solution is to think outside the box," she says.

There is also the matter of too much product. The solutions there will follow the same economics as in any other industry: consolidation or expansion. If revenues continue to decline, some producers will fall by the wayside or get swallowed up by competitors. As for expansion, there are plenty of countries that still actively ban porn. Getting a foothold in some of those now may reap huge rewards down the road. "The Chinese wall keeping pornography out is going to fall some day," says Kim Kysar of Pink Visual, which is trying to build its brand in neighboring Asian countries in anticipation of that eventuality. "As soon as it does, it's obviously going to be the biggest market there is."[39]

Nevertheless, many are still worried. *Hustler* president Michael Klein agrees with Tera Patrick. If the big porn producers can't figure out how to deal with piracy and the glut of product, they're going to have to figure out an entirely new way of doing business. "It's going to be harder to find the next Jenna Jameson or Tera Patrick, someone who's going to be a standout star," he says.[40]

The Final Frontier

The irony of the entire situation is that the destructive power of the Internet lies in the very seeds of its formation—the decision by Vint Cerf and DARPA in the seventies to freely distribute the rules that govern connections between computers. That seminal move was duplicated by Berners-Lee in the early nineties when he released his web browser code for free. Both actions, fundamental to the formation of the commercial Internet and Web, created the "culture of free" that has turned media business models—porn or otherwise—upside down over the past decade. The longer people use the Internet for free, the greater this sense of entitlement grows. Perhaps that is the way of the future and businesses will have to figure out a way of working within the new system, or perhaps old models will somehow reaffirm themselves, possibly through a continuation of lawsuits and enforcement of copyright laws. It will probably be a mixture of the two: content creators will eventually figure out how to do business and make money in the new paradigm, while users will accept that not everything on the Internet is automatically free.

There is, of course, one other possibility. Cerf is working with NASA's Jet Propulsion Laboratory in Los Angeles on developing the "intergalactic Internet," a new communications network that will connect all the satellites and spacecraft up in orbit with facilities here on Earth. The network got early DARPA funding in 2000, followed by a second round in 2007, then moved into production in 2009 after successful testing. Under the current system, if scientists on Earth want to retrieve data from, say, the Hubble Telescope, they need to schedule a download connection for when the sensor array passes over a node on Earth.

The new system follows the "hot potato" routing idea of the original ARPAnet, where data is stored at nodes and forwarded as soon as possible. Satellites and other craft will be able to automatically transmit their data to other nodes whenever they pass by them, eliminating the need for complicated logistical scheduling. This "delay-tolerant" feature, which sends information even if there are connection disruptions, also has applications here on the ground. The U.S. Marine Corps has tried it and loves it, as have Sami reindeer herders in northern

Sweden, who tested it for DARPA in 2005. Commercially, the system could theoretically be rolled out by Google on its Android cellphones for significant bandwidth savings. One Android user could, for example, download a map over the cellular network and then radiate it out to other Android users, thereby saving the other users from having to download it as well.

Given the history of the Internet so far, if the principles behind the "intergalactic Internet" are ever commercialized, there's little doubt that porn companies will once again be first in line to figure out how to capitalize on them. Who knows—maybe expanding porn sales to outer space is the solution to their problem of shrinking revenues here on Earth.

And how does the "father of the Internet" feel about the porn industry's role in nurturing and developing his creation? "It's slightly embarrassing to think that it might have grown up around that," Cerf says. "On the other hand, this is where VCRs did very well, too."

Outsourcing Burgers

Of course, it's not just the porn industry that is looking for ways to reimagine its business for the Internet era—fast-food companies are using it in increasingly inventive ways too. One of the most controversial by-products of the Internet has been the outsourcing of labor from developed to developing countries. Businesses of all stripes, from banks to phone companies to computer makers, have taken advantage of the low-cost communications made possible by the Internet to relocate their customer-service and IT operations to countries such as India, Brazil or the Philippines. After all, if a customer is calling for help from New York, it doesn't really matter on the service level if the company representative is in Houston or New Delhi. And it's much cheaper if the representative is in a developing country, where wages are a fraction of what they are in the United States or United Kingdom.

Outsourcing has been a double-edged sword, though. It has allowed companies to cut costs and create scores of jobs in places such as India. On the flip side, many customers resent having to deal with service agents on the other side of the world—they're seen as impersonal, unsympathetic and an example of how cheap the company employing them is. This discontent is largely psychological; even though Indian customer agents often provide the same level of service as their local counterparts, many Americans prefer to take their gripes to other Americans, while Australians simply want to complain to fellow Australians.

Big fast-food companies have been quick to jump on the outsourcing bandwagon. Through the early part of the new millennium, most moved in lock-step with other multinationals in transferring their customer service and IT operations to developing nations. But the fast-food chains did one better: they outsourced their drive-thrus. McDonald's, as usual, was the first to test the idea in 2004, moving order-taking from one state to another. How it works is simple: customers drive up and place their order as they normally would, but the person on the other end of the speaker is located hundreds or thousands of miles away. The connection is made using the same Internet-based calling technology that makes communicating with India so cheap and easy. The order-taker then relays the customer's order back to the restaurant, to be filled by employees there.

The new system, which has spread to many fast-food chains both in the United States and around the world, has several advantages over the old. It frees in-store employees from multitasking—rather than running around wearing a headset, taking multiple orders and then assembling those orders, workers can concentrate on one simple job, which reduces mistakes. It also speeds up the order-taking and processing of drive-thru customers. Wendy's, for one, has been able to shave between thirty seconds and a minute from customer order times. Given that fast-food restaurants make up to two-thirds of sales from their drive-thrus, those extra seconds translate into big dollars. In 2006 Wendy's said sales had jumped by 12 percent at stores testing the system. "It's the future of the industry. I can't believe how stupid I was not to do this sooner," said one Wendy's executive.[41]

One other benefit of the new system is the ability to move labor to less expensive jurisdictions. In the United States, this means outsourcing drive-thru order-taking from a state where the minimum wage or required benefits are relatively high to one where they are relatively low. The advantages of employing a call-center employee in Mississippi, where there is no minimum wage, for example, are obvious if your restaurant happens to be in, say, California, the birthplace of fast food, where the minimum wage in 2011 is $8 an hour. Taken to its next logical step, there seems to be little argument against fast-food companies taking the full plunge by outsourcing drive-thrus to call centers in developing countries, where they'll save even more on labor costs. Some, in fact, are already doing so—in 2006 one London curry restaurant outsourced its corporate take-out orders to India. "It is amusing that the orders travel nearly 5,000 miles to be finally delivered just half a mile away," the restaurant's owner said.[42]

Fast-food companies may prove once again to be on the

cutting edge by taking Internet-based outsourcing and applying it in new and innovative customer service applications. Bronco Communications, the company that managed McDonald's early outsourcing tests, is eyeing a way to expand its technology to retail chains such as The Home Depot, where customers would be equipped with Internet-enabled shopping carts. A call-center operator could then guide the customer through the store. A Bronco founder explains how this system would work: "You're at aisle D6. Let me walk you over to where you can find the sixteen-penny nails." Such possibilities, with their upsides and downsides, are virtually limitless. Fast food, meanwhile, will only get faster thanks to the Internet.

SEEDS OF CONFLICT

Monsanto's Genuity brand SmartStax corn has been bioengineered to produce its own insecticide and to resist herbicides.

World peace must be based on world plenty.[1]
—NOBEL LAUREATE JOHN BOYD ORR

George W. Bush is not likely to be remembered as a friend of science. In his eight years as president, he held back work on two of the most important scientific issues of our time: stem-cell research and global warming. In 2001, a month before the terrorist attacks consumed American domestic and foreign policy, Bush took to the radio airwaves to tell Americans of his fears regarding the use of cloned cells from humans in conducting medical research. The whole idea rankled with his ethical and religious sensibilities. "Researchers are telling us the next step could be to clone human beings to create individual designer stem cells, essentially to grow another you, to be available in case you need another heart or lung or liver. I strongly oppose human cloning, as do most Americans," he said in his address. "We recoil at the idea of growing human beings for spare body parts or creating life for our convenience . . . Even the most noble ends do not justify any means."[2]

With that, the president shut down federal funding into new lines of stem-cell research and sent a chill through the molecular biology community. Many researchers working in the field, appalled at Bush's misunderstanding and misrepresentation of their science, left the country to continue their work in more liberal environments, such as the United Kingdom. Stem-cell research was not about cloning people, they felt, but about solving some of the most crippling ailments that afflict humankind. Other nations were happy to accommodate these scientists and consequently took the lead in stem-cell research. In 2005 when South Korean scientists announced they had cloned human embryos (which turned out to be a hoax), Bush compounded his opposition in another speech: "I worry about a world in which cloning becomes accepted."[3]

Similarly, the president outraged scientists and environmentalists alike with his refusal to believe in global warming. While he paid lip service to fighting climate change in public, he also privately challenged the science behind it and refused to sign measures that would limit carbon emissions by American companies. Worse still, over the last few years of his presidency, details emerged of how scientists were silenced or forced to distort their findings by his administration. Researchers reported hundreds of instances of political interference in their work from 2002 to 2007, which they described as part of a concerted effort to confuse and obscure the global warming issue.[4] Bush's actions were summarized and roundly condemned by an editorial in the New York Times: "This administration long ago secured a special place in history for bending science to its political ends. One costly result is that this nation has lost seven years in a struggle in which time is not on anyone's side."[5]

What's curious is that during his time in office, Bush took a distinctly different view on another controversial scientific issue: genetically modified organisms (GMOs). While the idea of meddling with human genetics deeply troubled the president, rearranging the DNA of the things we eat was no problem. Indeed, Bush and the members of his administration were big supporters of GMOs and championed the cause around the world. In 2003 U.S. Trade Representative Robert Zoellick accused the European Union, with its "Luddite" opposition to genetically modified foods, of killing children in Africa: "I find it immoral that people are not being supplied with food to live in Africa because people have invented dangers about biotechnology."[6] Bush reiterated the sentiment in a speech a few months later and urged Europe to drop its ban on GMOs:

> We can greatly reduce the long-term problem of hunger in Africa by applying the latest developments of science. By widening the use of new high-yield bio-crops and unleashing the power of markets, we can dramatically increase agricultural productivity and feed more people across the continent. Yet our partners in Europe are impeding this effort. They have blocked all new bio-crops because of unfounded, unscientific fears. This has caused many African nations to avoid investing in biotechnologies, for fear their products will be shut out of European markets. European governments should join, not hinder, the great cause of ending hunger in Africa.[7]

Why the dramatically different attitude toward science as it pertained to genetically modified crops? Why was Bush's opinion on the merits of GMO technology drastically different from his beliefs on stem-cell research and global warming science? Some of the explanation can certainly be found in his religious beliefs, as well as his duty to promote American economic interests.

Some members of his administration also had previous links to the companies making genetically modified foods.

Human cloning provokes a number of ethical and moral questions; even many of the scientists engaged in the research have drawn moral lines on their studies. It's no surprise then, that Bush, a self-professed devout Christian, would find such research questionable at best. Global warming, however, is an issue with considerably fewer ethical dilemmas—everyone agrees it must be stopped and reversed, if possible. Many commentators have pegged Bush's opposition to fighting climate change to simple economic interests: American companies are polluting, but limiting their ability to do so would put them at a disadvantage with competitors in other countries such as China, which seem to care less about the environmental damage they're causing.

Pushing GMOs, on the other hand, also furthers the American economic agenda, but in a different way, because many of the technology's biggest developers, including Monsanto, Cargill and DuPont, are U.S.-based multinationals. After all, the American government has a responsibility to promote American companies around the world. Considering that food is a truly global business, it's understandable that Bush has been a big supporter of GMOs.

But there is another reason for the president's selective belief in science. September 11, 2001, came to define his presidency— virtually every policy decision he made after the attacks was tied to his crusade against terrorism. The "war on terror" consumed American domestic and foreign policy for pretty much the next seven years, from establishing the Department of Homeland Security to strengthening borders to dramatically boosting defense spending to invading Afghanistan and Iraq. Economic and scientific policies could not escape this mass mobilization any more than they could during the Second World War. Just as science was marshalled to fight the Axis powers, so too would it be used to combat the new threat of al Qaeda and other terrorist groups around the world, a struggle that has no apparent end in sight. Scientists and engineers in every discipline have been called on to help. As we'll see in later chapters, this has resulted in munitions experts creating new weapons, roboticists developing new forms of artificial intelligence and space exploration researchers coming up with new detection capabilities. As in every preceding American war, food scientists have had to do their part, too.

Genetically modified foods are another significant but subtle weapon in the war on terror. While missiles, tanks and guns aim to kill or incapacitate terrorists, GMOs are expected to work on a different level. They are supposed to ameliorate the circumstances that create terrorists and war in the first place. By creating and then farming crops that are resistant to drought, flooding, pests and other environmental threats, scientists can increase the amount of food available in poor countries. If those farmers can then use such crops to alleviate their food shortages, they can first feed themselves and then start on the road toward exporting, which holds the ultimate promise of economic prosperity. This idea, of using food to lift developing countries out of poverty, dispelling the hopelessness that leads people to take to war and terrorism, is our greatest hope for world peace. Or so GMO proponents like Bush want people to believe.

Better Green Than Red

The idea of food as a weapon is not a new one. For centuries, the strangling of food supplies has played an integral role in siege warfare, where a stronghold like a fortress or a city was encircled and starved into submission. While the idea is generally associated with medieval warfare, it finds itself alive and well into the twenty-first century—the decades-long American embargo of Cuba is just one example. (Cuba is also an example of a siege that simply has not worked.) For the most part, however, the use of food as a weapon has morphed into a subtler form over the past century.

As we've already seen, the Cold War manifested itself in several ways: the stockpiling of ever-more powerful nuclear weapons; proxy armed fights in places such as Angola, Chile and Vietnam; and indirect battle through the space race. The United States and the Soviet Union also competed fiercely to win over other, less powerful countries to their way of thinking. Besides steamrolling its way through Eastern Europe at the end of the Second World War, the Soviet Union fomented communist movements around the world in the fifties, from Africa to South America to Asia. The United States countered by financially and militarily propping up developing countries targeted by the Soviets. For the most part, the strategy worked in Latin America and Western Europe. In a few cases, however, the American government tried a better solution, one that involved providing developing countries with the means to become economically and militarily self-sustaining. Mexico was the first such experiment.

In 1940, when Manuel Ávila Camacho was elected president of Mexico, the country was facing food shortages and importing more than half the wheat it needed. Camacho, who hailed from a middle-class family and had some farming experience, was a proponent of industrialization and developing closer ties to the United States. Toward the end of the Second World War, he worked out a deal with the wealthy Rockefeller family to deploy new farming methods in Mexico. Both parties had a significant interest in keeping the population fed and happy. Camacho wanted to protect the country's fragile democracy while the Rockefellers had businesses, including oil operations, in Mexico that would have been jeopardized if the country fell to communism.

Norman Borlaug, a microbiologist from Iowa who had developed a host of chemicals for DuPont during the Second World War—including a disinfectant for canteens and a glue that held in salt water—was sent south in 1944 to work on the wheat problem. His solution, which took nearly ten years to put together, was a short, stocky hybrid plant, created through the cross-breeding of different kinds of wheat seeds. The new plant was resistant to the stem-rust disease common in Mexico, and with heavy watering and chemical fertilizers, produced significantly more grain than wheat grown from traditional seeds. The results were astonishing. Yields ballooned to the point where Mexico was a self-sufficient producer by the mid-fifties and a net exporter of wheat, shipping out half a million tons annually, by the mid-sixties.[8]

The securing of the food supply became one of the main pillars of the "Mexican Miracle," which along with investments in education and infrastructure allowed the country to post strong economic growth until the seventies. As one food historian put it, "a model for solving the stubborn dilemma of food insecurity seemed to have been hit upon, combining conventional genetics with the miracles of chemistry: Just add water and mix."[9] The "Miracle," which made Mexico a player in world markets and kept the majority of the population fed, helped fend off any thoughts of turning to communism as a solution for poverty.

With the Mexican success in hand, the American government exported Borlaug's farming technology to other countries. Next up were India and Pakistan, who in the mid-sixties were fighting famines caused by years of agricultural mismanagement. Borlaug's hybrid wheat seeds produced the same results as in Mexico. In India, wheat production increased nearly 80 percent in the first year of planting and just about doubled again in the second, with Pakistan seeing similar gains. By the mid-seventies, both countries were self-sufficient in wheat production.[10]

The idea of using genetically cross-bred seeds in conjunction with irrigation and chemical fertilizers was then applied to rice. A new hybrid seed, called IR8, was developed and planted in the Philippines. As with previous experiences, production doubled and promptly turned the country into a net exporter of rice. Borlaug was credited with saving millions of lives and awarded the Nobel Peace Prize in 1970 for the trend he had started, dubbed the "Green Revolution" because of the fields of green crops it produced. While a host of other factors, particularly financial and military support, helped Mexico, India, Pakistan and the Philippines fend off the seduction of Soviet communism, experts agreed that improved crop yields and the resultant economic self-sustainability played a large role. Borlaug himself alluded to how food mitigated conflict when accepting his award. "When the Nobel Peace Prize committee designated me the recipient of the 1970 award for my contribution to the 'Green Revolution,' they were in effect,

234 Sex, Bombs, and Burgers

I believe, selecting an individual to symbolize the vital role of agriculture and food production in a world that is hungry, both for bread and for peace."[11]

Wake Up, Sleepy Gene

The Green Revolution had its share of critics, not to mention some limitations that prevented it from being the silver bullet that ended world hunger. In places where it did work, the revolution was criticized for making farmers dependent on chemical fertilizers sold by the likes of Monsanto and DuPont. Continued use of pesticides also resulted in newer chemical-resistant weeds, which then escalated the need for newer, more powerful pesticides, and so on. Critics also said that by growing a single kind of wheat or rice rather than many different breeds, farmers ended up reducing the variety and therefore the nutritional quality of their diets, leading to malnutrition. Others said that the movement introduced non-sustainable farming practices such as over-irrigation, which caused ground water tables to dry up in places such as India, accelerating desertification.

The Green Revolution never took root in Africa because of the continent's general lack of water and varied soil qualities. While much of the movement's miracle results were due to technologically engineered hybrid plants and chemical pesticides, the seeds simply didn't amount to much without proper old-fashioned irrigation. Africa's soil quality also tended to differ within very short distances, which made it difficult to successfully plant and raise one single breed of crop. Detractors also argued that communism, not to mention proper capitalism, failed to take root because many African nations simply lacked the political infrastructure and stability needed to sustain any one form of government for very long. In Africa, the Green Revolution ultimately had a net-neutral geopolitical effect.

With the Green Revolution's limitations in mind, scientists continued their genetic experimentation on organisms. A huge breakthrough came in 1973, when a group of Stanford researchers transplanted genes from a frog into bacterial cells, marking the first successful attempt at creating "recombinant" DNA. The success sparked a debate within the scientific community about whether such research, which inevitably drew charges of playing God, should continue. A general consensus to move slowly was reached, but by the early eighties, that approach was out the window. In 1976 Herbert Boyer, one of the Stanford experimenters, founded the first biotechnology company, Genentech, to work on commercializing recombinant DNA. In 1982 the company launched an insulin product called Humulin, the first genetically engineered medicine to pass muster with the Food and Drug Administration. Humulin was synthesized in E. coli bacteria cells, which are commonly found in the human gut, then inserted into traditional insulin. The result was a longer-lasting drug that was better absorbed into the human bloodstream. Humulin proved to be a big hit and landed Genentech on the cover of *Business Week* and *Time*, kick-starting the biotech boom. As the eighties unfolded, a pharmaceutical revolution was under way, with genetically engineered drugs for everything from growth hormone deficiency to blood clotting hitting the market.

At the same time, research into creating better foods through the use of recombinant DNA was continuing. The first commercial product to pass FDA testing was the Flavr Savr tomato created by Calgene, a small company based in Davis,

California, northeast of the biotech hub of Stanford and San Francisco. Tomatoes had traditionally presented a problem for farmers because of their softness. In order to survive transport across hundreds or thousands of miles, they had to be picked while still green and firm, then artificially ripened with an ethylene gas spray at their end destination. Calgene scientists, however, managed to make a tomato that was resistant to rotting by adding a gene that fooled the fruit into ripening on the vine while at the same time retaining its firmness. The Flavr Savr was approved by the FDA for sale in 1994, but ultimately flopped because it delivered results opposite to those that GMOs promised. Calgene's tomato needed more land to grow than traditional breeds, which meant it was ultimately more expensive. Flavr Savr also didn't deliver superior taste or texture, so farmers and consumers couldn't really see the point. Struggling Calgene was eventually acquired in 1996 by Monsanto, which was quietly building its own genetically modified food empire.

Genetically Engineered Profits

Few companies have been as controversial for as long as Monsanto. The company was founded at the turn of the twentieth century by John Francis Queeny, a veteran of the then-fledgling pharmaceutical industry, and its first major customer was Coca-Cola. Besides selling the soft-drink company the artificial sweetener saccharin, Monsanto also turned Coke on to caffeine, which was added into its soft drinks. By the forties, Monsanto had grown into a multinational company and expanded into plastics and other chemicals. Like virtually every American business at the time, the company joined the fight against Germany and Japan and helped produce chemicals for the Manhattan Project. After the Second World War, it became a big producer of the pesticide DDT, which positioned the company nicely to create a defoliant for American forces in the Vietnam War. That chemical, Agent Orange, turned out to be as dangerous as DDT. It was found to be highly carcinogenic to anyone who came in contact with it, which ultimately resulted in Monsanto paying out hundreds of millions of dollars to victims in lawsuits and settlements. The company has also been busted in several countries for illegally dumping toxic waste in lakes and rivers. Given its size and historical dominance of the chemical market, Monsanto has over the years become the environmental movement's Public Enemy No. 1. It is equally loathed as the quintessential Big Corporation that lobbies, sues and bullies its way into getting what it wants. A cottage industry of books and documentaries has sprung up to criticize Monsanto, making it one of the most reviled American companies this side of Halliburton, McDonald's, and Wal-Mart.

Monsanto led the charge in the Green Revolution and, subsequently, the move toward genetically modified organisms. The company developed and patented a herbicide called glyphosate and began selling it in 1973 under the brand name Roundup. Farmers couldn't get enough of the chemical which, to paraphrase the marketing line for Raid bug spray, "kills weeds dead." By the early eighties, Roundup was the top-selling herbicide in the world.

But as every technology company knows, you're only as good as your latest product, especially when your patent eventually expires (in the United States it's seventeen years). Monsanto's next product was particularly ingenious. To keep farmers from shifting to cheaper, generic imitations of Roundup after its patent expired, the company developed genetically modified seeds that worked in conjunction with its herbicide—and *only* its herbicide. Roundup Ready soybeans became available in 1996, followed in 1998 by Roundup Ready corn/maize, and soon thereafter with canola/rapeseed and cotton. The seeds were genetically engineered to resist Roundup, so the farmer could spray the herbicide onto an entire field rather than having to selectively find and eliminate the weeds one by one, a huge time savings. In technological parlance, Monsanto had created a "white list"—only those plants it deemed worthy would survive its chemical killer. In business terms, the company had created a market advantage that could not be matched by competitors once its patent on Roundup expired.

The company also perfected a drug called Hygetropin that boosted milk output from cows, which it put on the market in 1994 as Posilac. Hygetropin was created in similar fashion to Genentech's Humulin, by inserting a cow's natural growth hormones into E. coli bacteria, where they were separated into a purer form and then injected back into the cow. The result, Monsanto said, boosted a cow's milk output by about 10 percent a year. The company also perfected insect-resistant corn and cotton seeds by injecting them with genes from the B. *thuringiensis* bacteria. The resultant Bt corn and Bt cotton plants essentially secreted the bacteria, which is harmless to humans but deadly to insects looking for a snack.

While Monsanto became the biggest and most active researcher and purveyor of genetically modified organisms through the eighties and nineties, it was hardly the only one. Other big chemical companies, including American duo DuPont and Cargill, Britain's Zeneca, France's Aventis and Belgium's Plant Genetic Systems, all moved to get a piece of the burgeoning biotech pie, which in the mid-nineties represented billions upon billions of dollars in potential profits. The road to those riches, however, was anything but smooth.

Cow Licked

The outcry over GMOs started not with a bang but with a moo. In the late eighties, Britain discovered its cattle had fallen prey to Bovine Spongiform Encephalopathy (BSE), better known as mad cow disease. The sickness, which slowly turned the cows' brains to mush, was found to have been caused by feeding cattle a deadly cocktail of infected protein supplements and the remains of other infected cattle. Considering that cows normally eat grass, it's not much of a surprise that they got sick off the bovine equivalent of Soylent Green. In the early nineties, as the disease was discovered to be widespread and the panic rose, other European countries banned British beef. British food regulators, however, maintained that the disease was limited to cows and had absolutely no effects on humans. How wrong they were. In 1996, just as cans of tomato paste made from Flavr Savr were starting to creep into Britain, scientists linked BSE to a variant of Creutzfeldt-Jacob Disease, which destroys brain tissue in humans just as BSE does in cows. Eating contaminated beef meant you could contract a fatal brain disease, and cases began to pour in. As of early 2009, 165 people had died in Britain and 23 in France, with more cases expected, since the disease can take up to forty years to incubate.[12]

The manure hit the fan. Public outrage at being misled by regulators was palpable and manifested in the outright rejection of the chemical cocktails cattle had been pumped with. The anger spread to all of the supposedly "safe" technologies used in food production and inevitably targeted the new genetically engineered foods, which came along at exactly the wrong time. The anti-GMO crusade was led by a few high-profile names, including Greenpeace and Prince Charles. Greenpeace blockaded European ports where American GMO shipments were due to arrive while the heir to the British throne stoked fears through the media. In a front-page article in the *Daily Mail* in the summer of 1999, the prince posed ten questions such as "Do we need GM food in this country?" and "Is GM food safe for us to eat?" His answer to both rhetorical queries was an emphatic "no." Genetically modified foods would lead to an "an Orwellian future" and "the industrialization of life itself," Charles wrote.[13]

European grocery stores responded first by labeling foods that contained genetically modified ingredients. When nobody bought them, they were eliminated entirely. Restaurants, including fast-food chains, forbade suppliers from using GMOs for fear of public backlash. The furor then spread beyond Europe. In North America, image-conscious food companies worried about the potential damage if they were targeted by anti-GMO crusaders and quietly moved to limit their exposure. In 2000 Simplot, the potato people, ditched Monsanto's New Leaf spud, which like the company's Bt corn and cotton was genetically engineered to produce its own pesticide, after getting its marching orders from McDonald's. "Virtually all the [fast-food] chains have told us they prefer to take non-genetically modified potatoes," a Simplot spokesman said.[14] Monsanto gave up on New Leaf a year later, once potato-chip makers Frito-Lay and Procter & Gamble joined the boycott. Monsanto was also forced to abandon Posilac, its growth hormone for cows, after critics charged it accelerated mastitis—an inflammation of breast tissue that resulted in, among other things, pus-filled milk—and also cancer in humans. Regulators in Japan, Australia, New Zealand and Canada, a country that had typically been very friendly to GMOs, refused to allow the sale of genetically engineered cow growth hormone. Monsanto finally unloaded what critics have called "the most hated product in the world" to drug maker Eli Lilly in 2008. The transaction wasn't the end of Posilac, however, with Eli Lilly announcing it would "continue to provide dairy farmers with this important production tool."[15]

The European furor also had a significant impact in Africa. Fearing their products would be shut out of important European markets, governments and farmers refused to grow genetically modified crops. A number of countries, including Nigeria, Sudan, Angola, Zimbabwe, Namibia, Mozambique and Malawi, went a step further and refused to accept any food aid that contained GMOs for fear that such products could contaminate their own crops. Governments in those countries decided they would rather let their people starve than risk contaminating their food supplies with unknown substances. The "Luddite" and child-killing comments that Bush and his administration leveled at Europe were hardly surprising.

You Can't Stop Progress or GMOs

The controversy has certainly not stopped the inexorable march of genetically modified foods. In North America, where regulations governing their use are virtually non-existent, they're everywhere. The landmark ruling came in 1992, when President Bush (senior) decreed that GMOs were not substantially different from traditionally grown crops and therefore required no special labeling. Canada, which generally rubber-stamps American health rulings, followed suit. Now, about two-thirds of the processed foods found on North American grocery store shelves contain genetically modified ingredients.[16] Produce aisles are naturally full of such foods, as are meat departments, since livestock is fed on GMO crops such as corn. Restaurants can't help but serve them to customers and even fast-food chains such as McDonald's, which rejected Monsanto potatoes and instituted a widespread ban of GMOs in Europe, are using them in North America.

For Americans and Canadians, it's virtually impossible to get around consuming genetically modified organisms even if they want to, because the technology is in just about everything and there are no laws requiring labeling. Indeed, food retailers who have tried to label their foods as GM-free, such as the Ben & Jerry's ice cream chain, have been sued into silence by the likes of Monsanto, who have argued that identifying foods as such implies their superiority to those containing GMOs. The administrative attitude toward genetically modified foods in North America is unlikely to change under President Barack Obama, even though he wasted no time in reversing many of his predecessor's science policies, including the funding ban on stem-cell research. The main backing for GMOs may have come from the two Bush presidents, but Obama continued the trend when he named Iowa attorney and GM supporter Tom Vilsack as his Secretary of Agriculture. Obama, however, has at least indicated that he is interested in hearing the other side of the story by appointing organic food expert Kathleen Merrigan as Vilsack's deputy. That decision may have been influenced by Obama's wife, Michelle, who is an avowed fan of organic farming. The First Lady sent the biotech industry into a tizzy in early 2009 when she planted an organic garden at the White House.

In terms of production, worldwide growth is quickly accelerating. In 2008 biotech crops globally accounted for 480,000 square miles, up nearly 10 percent from the year before. Total crops planted between 1996 and 2008 reached two billion acres, an impressive feat given that it took a full ten years to reach the first billion, but only three years to get to the second billion. Total acreage is predicted to at least double again by 2015. The number of countries planting GMOs reached twenty-five in 2008, with fifteen of those—including Colombia, Honduras and a pair of African countries in Burkina Faso and Egypt—classified as developing nations.[17] The United States leads the way in total production, making up nearly 60 percent of the world's GMO output, followed by Argentina at 20 percent, Canada and Brazil at 6 percent each, and China at 5 percent.[18] The total number of countries growing GM crops is expected to grow to forty by 2015.

Soybeans are the most popular crop grown, accounting for more than half of the area currently planted with GM foods worldwide, followed by corn, cotton and canola. A few countries recently added new crops to that cohort, with the United States now growing GM squash, papaya, alfalfa and sugar beets, while China is farming tomatoes and sweet peppers. Genetically modified versions of the world's two biggest crops, wheat and rice, have been created and are in various states of regulatory review. Scientists have also successfully tested drought-resistant crops, as well as seeds that "stack" genetically modified traits. SmartStax corn, which Monsanto introduced in 2009, combines the abilities of its Roundup Ready and Bt products in that the plants are resistant to the company's herbicide and secrete their own pesticides.

Genetically modified foods produced from animals are also on their way. The FDA in 2010 said it was close to approving AquaAdvantage, a salmon that has been genetically modified to grow faster, for human consumption. If and when the fish gets the green light, other GM animals will surely follow.

Despite the spread, resistance is still firm—and entrenched in parts of Europe. In 2000 the European Union believed it had won a victory with an international agreement that allowed signatory countries to monitor and test GMOs for potential environmental effects before they approved them for commercial use. The Cartagena Protocol, which came into force in 2003, was flawed, however, in that it could be trumped by World Trade Organization agreements. The United States wasted no time in taking its case to the WTO and won a historic ruling against the EU in 2006. Under the decision, European countries could not obstruct the entry of genetically modified products if there was demand for them within their borders.

GM foods have begun to creep into Europe, but the fight continues as individual countries have recently resorted to the old European standby argument of sovereign rights. Austria, Hungary, Greece, France and Germany have all declared GMO bans of various levels, as have non–EU members Switzerland and Albania. American interests quickly responded with that old American standby, the lawsuit. Monsanto jumped on Germany in April 2009, just days after the government announced a ban on the company's corn. "They are in conflict with EU rules," a company spokesperson said.[19]

The anti-GMO rhetoric also continues. In 2008 Prince Charles reiterated the concerns he voiced nearly a decade earlier. Companies such as Monsanto are conducting "a gigantic experiment, I think, with nature and the whole of humanity which has gone seriously wrong," he told *The Daily Telegraph*. Relying on large corporations for food would result in "absolute disaster" and the "destruction of everything."

"If they think it's somehow going to work because they are going to have one form of clever genetic engineering after another then again, count me out because that will be guaranteed to cause the biggest disaster environmentally of all time."[20]

Kill Your Enemy, Fill Your Belly

The rhetoric on the other side of the argument is also ratcheting up. While American corporate interests have been and continue to be a big driver of GMOs, an increasing number of scientists, both social and biological, are voicing their support as well. While GMO critics argue that the world has enough food and that it simply isn't being distributed correctly to the people who need it, many social scientists disagree. The world may actually be heading for disaster because of rampant population growth. Over the past half-century, the world's population has grown more than it did during the previous *four million* years and is expected to double again over the next fifty years.[21] Just about all of that growth is expected to happen in the developing world, where 800 million people already have insufficient food.[22] At the same time, arable land is decreasing at the rate of about 1.5 percent a year because of the three deadly "ations"—desertification, salinization and urbanization.[23] China and India, the world's two most populous nations, are already near a crisis point, each using about three-quarters of its available farmland.[24]

These facts set the stage for what indeed could be, to use Prince Charles's words in a different context, "absolute disaster."

There's a concept, called the population-national security theory, that was postulated by social scientists during the Green Revolution and that neatly sums the situation up. It goes like this: a growing population results in overcrowding and exhaustion of resources, which in turn leads to hunger and political instability. Political instability then leads to communist insurrection, which is a danger to American interests. And what's the ultimate result of threatening American interests? In many cases, it's been war. President Harry Truman vouched for this theory in his inaugural address in 1949. "More than half the people of the world are living in conditions approaching misery. Their food is inadequate . . . Their poverty is a handicap and a threat both to them and to more prosperous areas," he said. "Greater production is the key to prosperity and peace. And the key to greater production is a wider and more vigorous application of modern scientific and technical knowledge."[25]

The theory has been repurposed for modern times, with the word "terrorist" replacing "communist," and it has many supporters within government and the scientific community. Right up until his death in 2009, Norman Borlaug backed the use of genetically modified crops as a tool to boost food production levels and fight the conditions that create war and terrorism in developing countries. "This is the most fertile ground for planting all kinds of extremism, including terrorism. And the people of the developed nations won't live in peace and tranquility with that pot boiling over," he said. "First, it's internal conflict in a country, civil war. Then other countries get involved and here we go again. Those are the dangers."

Borlaug, who in addition to the Nobel Peace Prize was awarded a litany of accolades, including the Presidential Medal of Freedom, the Congressional Gold Medal, the National Medal of Science and the Padma Vibhushan (India's highest honor to non-citizens), carried considerable weight in the debate. Having already been credited with saving more than 240 million people from starvation, he continued to campaign on into his nineties for the cause of using technology to solve hunger.[26] He even appeared in Monsanto promotional videos to defend the company's genetically engineered crops. "What we need is courage by the leaders of those countries where farmers still have no choice but to use older and less effective methods," he said in one video. "The Green Revolution and now plant biotechnology are helping meet the growing demand for food production while preserving our environment for future generations."[27]

But how much of a factor is hunger in driving people to take up arms? While a number of inputs, including politics, religion and simple aggression all contribute, social scientists and war historians agree that poverty, hunger and the hopelessness they create are among the biggest motivators. Peter Singer, a social scientist at the Brookings Institution in Washington, D.C., and author of several books on war, says it's no different from why people turn to crime. "You have people go into crime because of desperation or they go into it because of the context of how they were raised where poverty was a driver. You have people go into crime because they're greedy or they're downright evil and they were born that way. Much of the same parallel can be made as to why conflicts and wars start." Poverty and desperation tend to hollow out the social and political institutions that are needed for good governance and for stability and prosperity, he says. "Conflict entrepreneurs take advantage of that absence of good governance."[28]

Food and the escape from poverty are among the key drivers of recruitment for conflicts in Africa, particularly among children. As many as 250 million children live on the street, more than 210 million must work to feed themselves and their families and one-third of all children suffer from severe hunger. In *Children at War*, which looks at the horrific rise of child soldiers over the past few decades, Singer found that such hopelessness presents a huge pool of labor for the illegal economy, be it organized crime or armed conflicts.[29]

The children themselves point to food as a major reason for why they enlisted as soldiers. Fighting may be a dangerous choice of profession, but in many cases it's better than the alternative. "I don't know where my father and mother are. I had nothing to eat. I joined the gunmen to get food," said one twelve-year-old soldier in the Congo.[30] "If I left the village I would get killed by the rebels who would think I was a spy. On the other hand if I stayed in the village and refused to join the army, I wouldn't be given food and would eventually be thrown out, which was as good as being dead," said another, aged fourteen. "I heard that the rebels at least were eating, so I joined them," said yet another.[31]

The story is similar in parts of the Middle East. In Afghanistan, where decades of war have destroyed virtually every institution, hunger is rampant. A pair of Afghan boys told Singer they had the choice of following a cow around to scoop up its excrement to sell as fuel or joining one of the armed factions. Enlisting provided them with clothes, food and a shred of self-respect.[32] Graeme Smith, a Canadian journalist who covered the war there for three years for *The Globe and Mail* says such stories are numerous. The reasons Afghans enlist with al Qaeda and the Taliban are usually not political or religious, as the Western media would have us believe. "They're inextricably linked, hunger and war. Right now, hunger is absolutely one of the big factors driving the conflict," he says.[33]

Since the American invasion in 2001, fighting and the resultant deaths have followed weather and agricultural patterns. Fighting usually kicks off after the country's main cash crop, the poppies from which opium is derived, is harvested in the spring months and continues until it gets cold in December. Day laborers tend to get paid well during the poppy harvest, Smith says, but after that they are at a loss for ways to buy food. For many—like the cow poop boys—enlisting is the only option. As Charles Stith, the former U.S. ambassador to Tanzania, puts it, this is fertile ground for terrorist organizations on recruiting drives. "The foot soldiers of terrorist groups tend to be on the lower rungs of the socio-economic ladder," he says. "People who have hope tend not to be inclined to strap 100 pounds of explosives on their bodies and go into a crowd and blow themselves up. People who have hope are not inclined to lie in wait outside an airport with a missile looking for a plane full of tourists to shoot down."[34]

Recruiting the poor, hungry and hopeless isn't a tactic reserved for terrorists and African warlords. It's also a long-standing practice in developed nations, although the idea of poverty is relative in such places. Since food—especially the unhealthy, heavily processed kind—is plentiful and cheap in prosperous countries, recruitment of the poor usually takes advantage of a person's lack of education or job prospects. The U.S. government, for one, has had to deal with charges of using a "poverty draft" for its forces ever since the end of mandatory conscription following the Vietnam War, despite pitching military service as a good way for recruits to earn college tuition. During the first Gulf War, African-American leaders criticized the disproportionate numbers of blacks in the military compared to whites and the population in general. African-Americans, usually from economically depressed areas of the country, made up a quarter of all troops in Iraq in 1991, but only 12 percent of the population. One study found that 33 to 35 percent of all qualified black men at the time had served in the military, more than double the percentage for white men. "This nation ought to be ashamed that the best and brightest of our youth don't volunteer because they love it so well, but because this nation can't provide them jobs," said Benjamin Hooks, the president of the National Association for the Advancement of Colored People.[35]

Fifteen years later, little had changed for the second Iraq war. Concerned citizens in New York, in one example, began organizing "counter-recruitment" rallies in the face of what they saw as increasingly aggressive attempts by military recruiters to draw in students from high schools in poor areas such as East Harlem. Barbara Harris, one of the protesters, passed information on to students about the financial aid they could receive for college. "If a young person wants to enlist, at least he or she knows what it's about, what the truth about recruiting is. They can decide if that's the best choice for them."[36]

Nevertheless, the United States has moved quickly to try to fight hunger as a conflict motivator in Iraq, particularly with GMOs. In 2004 the U.S.-led Coalitional Provisional Authority government, in place since the 2003 invasion, handed control back to Iraq's own government and issued its controversial 100 orders. The rules were designed to transform the country from a centrally planned economy to a market-driven one, but critics suggested they were really intended to facilitate a form of American economic colonialism. Order 81—which sounds like the ominous directive given by the evil Emperor to exterminate the Jedi in the *Star Wars* movies—clearly opened the door for American GMO producers. The order's Plant Variety Protection clause allows for the patenting of new plant forms, or genetically modified crops. Iraq's agricultural system was badly shattered during the first Gulf War and wasn't allowed to fully recover because of American and British sanctions afterward, but it is still in better shape than Afghanistan's. With the worldwide Islamic Jurisprudence Council having approved GMOs for consumption in 2000, it will only be a matter of time before Iraqi farmers are awash in Roundup Ready products. In Afghanistan, where the agriculture system is little better than it was in the Stone Age, it'll be a while yet.

Patenting Humanitarianism

But the enemies of GMOs don't buy the "make food not war" argument or the promises of humanitarianism put forward by purveyors. While Prince Charles has called the playing of the Africa card "emotional blackmail," Greenpeace continues to maintain that the only people who benefit from genetically modified foods are the shareholders of the large biotech companies. The proof is in the pudding, says Greenpeace Canada's anti-GMO campaigner, Eric Darier. The technology to provide drought-resistant or nutrient-enhanced crops is possible, but the only seeds to have been commercialized since the mid-nineties are those tied to chemical fertilizers. "There's nothing new per se. There were a lot of promises and we at Greenpeace were saying that wasn't the purpose. The purpose was for Monsanto to control the seed market and to be able to push their own herbicides," he says. "It's a very sophisticated way of controlling the market."[37]

Some farmers who plant Monsanto seeds, both in North America and in India, have in fact complained about the company's "technology user agreements," which contain a number of restrictive clauses. One such limitation, for example, prevents farmers from saving seeds from year to year. Critics say this clause is intended to force farmers to buy new seeds every year, but Monsanto insists it's because GM products, just like Norman Borlaug's hybrid seeds before them, don't reproduce very well.

The other major impediment to humanitarian uses of GMOs, critics say, is the actual patenting of the seeds themselves. In 1999 a German plant science professor named Ingo Potrykus, the biotech incarnation of Borlaug, came up with a seed he called Golden Rice while working at the Swiss Federal Institute of Technology.[38] The genetically engineered rice, which is yellow or orange in color, produced significantly higher levels of vitamin A and was positioned to solve one of the biggest malnutrition problems in the world. The World Health Organization says up to 250 million preschool children in 118 countries suffer from vitamin A deficiency, which can lead to blindness and ultimately death.[39] Researchers at the Institute of Food Technologists peg the number of deaths per year at between one and two million.[40] Suddenly, as the new century approached, the potential to use the newest food technology to save a significant number of lives looked like it would finally be realized.

But Potrykus found that, despite the fact that he had created his rice in an academic setting free from corporate influence, the issue of intellectual property still crept up. Not only had Monsanto and other companies patented all their seeds, they had also protected the techniques used to make them. Potrykus's Golden Rice, it turned out, was unknowingly in potential violation of seventy different intellectual and technical property rights held by thirty-two different companies. If the rice were to be disseminated to poor farmers, each of those rights would have to be negotiated, a fact Potrykus found deplorable:

> It seemed to me unacceptable, even immoral, that an achievement based on research in a public institution and with exclusively public funding, and designed for a humanitarian purpose, was in the hands of those who had patented enabling technology early enough or had sneaked in a material transfer agreement in the context of an earlier experiment. It turned out that whatever public research one was doing, it was all in the hands of industry (and some universities).[41]

Potrykus soon changed his tune, though, when AstraZeneca, the large Anglo-Swedish pharmaceutical company that held much of the intellectual property used in Golden Rice, offered to negotiate a deal that would allow farmers free access to the patents. Potrykus then turned his criticism on the various regulatory agencies around the world, particularly in Europe, that demanded his rice jump through all sorts of hoops to get approval.

Ten years after it was invented, Golden Rice was still not available anywhere in the world. Food scientists who back GMOs are livid that such strict regulatory review is being enforced when so many people are dying. It's a situation that illustrates just how emotional and paranoid people in the developed world have become about food, a luxury that people in the developing world don't have. Bruce Chassy, the associate director of the biotechnology center at the University of Illinois, says products like GMOs should receive the same sort of regulatory fast-tracking that drugs like AIDS medications get. "We can't spend thirty months monitoring a drug while people are dying," he says. "I have a problem with this moral equation. What is it about one to two million people dying a year from vitamin A deficiency that doesn't make you want to try out just about anything?"

Golden Rice finally went into field tests in the Philippines in 2008 and may become commercially available to farmers there in 2011, but its long road to market highlights the problems of using GMOs for humanitarian purposes. Critics say the issue of patents slows down and discourages research into non-profit-based uses of GMOs, while advocates argue it's the overly cautious approach of regulators, influenced by the emotionally charged scaremongering of critics, that is impeding progress.

Ultimately, barring a huge disaster, GMOs will continue their spread. With the continued growth in population, food will become scarcer, which means that conflict—war and terrorism—will likely only increase. As Chassy puts it: "There's this giant train barrelling down the tracks at us and it's going to cause more civil unrest and suffering in the world than anything conceivable."

If we think there's a lot of conflict in the world today, we ain't seen nothing yet. If GMOs are indeed, as Prince Charles says, a giant experiment, it may just be an experiment worth trying.

FULLY FUNCTIONAL ROBOTS

The Packbot from iRobot, which is used in reconnaissance and bomb disposal in Iraq and Afghanistan, uses a PlayStation controller to direct its movements.

People are willing to have sex with inflatable dolls, so initially anything that moves will be an improvement.[1]
—EUROPEAN ROBOTICS NETWORK CHAIRMAN HENRIK CHRISTENSEN

There aren't a lot of benefits to being a journalist. The pay isn't great, there's the constant stress of deadlines and people are always indirectly blaming you, "the media," for sensationalism, blowing things out of proportion or, my favorite, reporting something "out of context," the default excuse of people caught saying something they shouldn't have. There are a few bright sides, though. We tend to get a lot of free coffee and sandwiches, and every now and then we get to interview one of our childhood heroes. (Kung fu action star Jackie Chan comes to mind.) On the rarest of occasions we see or experience something that completely blows our mind and makes up for all the bad coffee.

That's what happened to me in January 2008, while I was covering the Consumer Electronics Show in Las Vegas. After wading through the crushing crowds and the sensory overload that is the convention floor, I staggered out to the parking lot for my appointment with the "Boss," the robot vehicle built in Pittsburgh by Carnegie Mellon University engineers. Just two months earlier the souped-up General Motors SUV had won the DARPA Urban Challenge, in which fully automated vehicles raced around a 60-mile track in a simulated city environment.

After explaining how the vehicle worked—it used a combination of radar, laser sensors, cameras and GPS positioning—project manager Chris Urmson took me for a ride around the obstacle course set up in the parking lot.

With me in the passenger seat and Urmson in the back, the Boss lurched to life and began to drive the oval-shaped course, deftly veering around the garbage can and pylon obstacles. I sat there open-mouthed, staring in awe at the empty driver's seat and the steering wheel as it eerily turned itself left, then right, then left again. I had a flashback to *Knight Rider*, the eighties show in which David Hasselhoff drove an intelligent car named KITT. (The tricked-out Trans Am could drive itself, have conversations with people and even, in one outrageously silly episode, help Hasselhoff gamble by somehow magically controlling his dice.) Here was KITT in reality. My mind reeled at what was happening, and what it meant. Sure, I'd seen robots on TV and even a few simple ones working in person, yet here one was *chauffeuring* me around.

After a few laps around the course, the Boss unexpectedly veered left, jerking me out of my reverie. The car plowed into some garbage cans set up on the side of the obstacle course, then came to an abrupt stop. "That's never happened before!" Urmson exclaimed from the back seat. Having seen too many movies, I immediately started thinking the machine was turning on us, as in *I, Robot* or *The Matrix*. A few tense seconds passed as I tried to remember how Sarah Connor had defeated Arnold Schwarzenegger in *The Terminator*, but then the Boss came back to life. The car calmly backed up and resumed its course. The test drive ended a few minutes later and Urmson set to figuring out what had happened. It turned out the Boss's cameras weren't able to see the lane markings on that particular patch of the course, hence the swerve. There was no harm or damage, but still, I thought, robot cars evidently have some way to go before they can be trusted on the streets.

The process will indeed take some time as robotic features are added one at a time, Urmson said. Some GM vehicles already incorporate technology used in the Boss, such as lane and blind-spot detectors that alert drivers when they are swerving. Next up will be a form of robotic cruise control where the car can detect how fast or slow the vehicles in front of it are going, then adjust its own speed accordingly. Eventually, self-driving cars will be allowed on highways, since traffic there is more straightforward than on city streets, and drivers will be freed up to do other things on long-distance journeys. "It's going to be phased in gradually but we expect a fully autonomous, self-driving car to be on the road in the next decade," Urmson said. "That means during a long road trip, I can read, watch a movie or even sleep." A GM executive listening in on our conversation also suggested that once the bugs are worked out, robot drivers will be safer than humans because they have no emotions. "There will be no more road rage because it's logical. It's like Mr. Spock."[2]

The IBM of Robots?

The reality of the impending revolution hit me that day in Las Vegas. Robots are no longer science fiction; they aren't just automated arms in car factories or cutesy toy dogs anymore. They are here, they work (mostly) and they will soon be everywhere. Self-driving cars will be only one small phase of the coming wave. The global robotics market, made up mostly of those automated arms, was pegged at $17.3 billion in 2008 and is expected to grow massively over the next decade as new uses take off—up to $100 billion by one estimate.[3] There are now an estimated 55 million robots in homes around the world in the form of toys, vacuum cleaners, lawn mowers and security monitors, and some people believe it won't be long before every household has one. Indeed, the government of South Korea has mandated such a plan to be in effect by 2020. Robots will soon feed us, clothe us, wash us, keep us company and fetch us beer from the fridge. Eventually, we'll even have robots to control those robots. Many look at the industry's ramping growth and compare it to the early personal computer market of the seventies. "We may be on the verge of a new era, when the PC will get up off the desktop and allow us to see, hear, touch and manipulate objects in places where we are not physically present," says Microsoft founder Bill Gates.[4]

But when will this revolution happen, and who will lead it? Tandy Trower, general manager of Microsoft's robotics department, believes the new era will be ushered in by a big carmaker, namely Toyota. In an industry where a single leading company has yet to emerge the way IBM did with early computers or Microsoft with software, the Japanese carmaker has a number of advantages over other robotics manufacturers, Trower says. Toyota has decades of experience in robotics, since it was one of the first to make wide-scale use of automated production lines. The company has also produced some amazing examples of advanced robotics, such as robots that can play the trumpet and violin. Toyota also understands the consumer market, has deep pockets for the heavy research and development needed and a distribution network to make mass adoption a reality.

Toyota has taken a smarter approach than most American companies, which have been too fixated on the military market,

Trower says. The biggest market for robotics lies in health-care and personal assistance, which is where the car company is looking. "Rather than focusing on the military applications, they're focused on the social aspect of how robots will assist us in the future. They are a very important company to watch," he says. "Toyota could be the IBM of robots."[5]

I don't agree because history argues the contrary. Health care and elderly assistance may indeed prove to be the biggest home market for robotics makers, but it certainly won't be the first large one. If our story so far has taught us anything, it's that the early adoption of this kind of technology is almost always spurred by two of our most elemental human behaviors: sex and violence. Toyota has few, if any, links to the industries serving those two markets. Second, while a Japanese company may lead the first wave of mass adoption, it is not likely to be a big one like Toyota. Again, history has shown that big companies rarely lead the spread of new technologies because their very success is tied to existing or "old" technologies. Investing too much in new advances risks the possibility of spreading a company thin or cannibalizing an existing business, which is exactly what has happened to Trower's own employer, Microsoft, in recent years. Because the software giant made its fortune supplying the operating system on which desktop computers run, it was slow to address the rise of the Internet, where it doesn't really matter which operating system you're running. Microsoft squandered its position of dominance on the computer to Google, a company built from the ground up *on* the Internet. As such, Google now dominates the main form of revenue generation on the Internet—search-related advertising—

the way that Microsoft dominates computer operating systems. The only difference is that one business represents the future while the other is shrinking in importance.

The same phenomenon is likely to happen in robotics. While a company like Toyota maintains a significant advantage in terms of resources and experience over the thousands of tiny robotics start-ups around the world, its main business is still selling cars. Toyota's shareholders may think robotics is quaint and interesting, but their main concern will always be selling cars. They are likely to resist any big incursions into businesses that have little to do with core competency. Small robot companies don't have such an existing business to protect and are much more likely to seek out and address opportunities wherever they present themselves. One such company, iRobot, is already emerging, and for my money it is far more likely to be the Microsoft or IBM of robots.

Robots That Suck

Based in Bedford, Massachusetts, a short drive north from Boston through leafy New England, iRobot's headquarters is a single building in an industrial park overlooking the freeway. It's a much smaller base than what you would expect for what is quickly becoming one of the world's most important companies. Still, the modest operation speaks volumes about the company and the industry in general—while it is full of promise and has had some success so far, it is still very much a nascent business. iRobot was founded in 1990 by Colin Angle, Rodney Brooks and Helen Greiner, a trio of robotics researchers from MIT, and took its name from the Isaac Asimov novel *I, Robot*—since turned into a movie starring Will Smith—wherein humans and machines live together in relative harmony (at least before the machines rise up and rebel). In its early days, the company built some impressive robots, including a toy dinosaur, but floundered about in search of a market for its technology. "Probably about fourteen or eighteen business models came along and were discarded as they were found to be little more than a subsistence existence," explains Angle, the chief executive officer.[6] Like many high-tech executives, Angle eschews a suit in favor of casual attire, a simple button-up polo shirt to complement a slightly disheveled look that he seems to have borrowed from Bill Gates.

Opportunity finally came knocking in 1997 when the company designed Fetch, a robot that cleaned up cluster bomb shards from airfields, for the air force. That led to a DARPA contract the following year for the PackBot, a robot that resembles a lawn mower, but with tracks instead of wheels and a long multi-jointed arm sticking out of it. The "platform," as iRobot calls it, can be customized with whatever equipment is desired. The arm can be equipped with several different types of cameras, including night-vision, as well as additional claw-like "hands" or even sensors for detecting explosives and biological weapons. Weighing 45 pounds and costing $150,000, the PackBot turned out to be a godsend for troops in Afghanistan and Iraq. Its small size and ruggedness allowed it to go just about anywhere, from caves to office buildings, and its customization options let it perform many different tasks, including reconnaissance and bomb disposal, which has come in handy battling the weapon of choice of Iraqi insurgents, the improvised explosive device (IED).

The military contracts gave iRobot financial stability and allowed Angle and his cohorts to think about their real goal: the consumer market. In 2002 the company rejigged the Fetch into the Roomba, a disc-shaped vacuum cleaner that looks like a big Frisbee. At $200, the Roomba was the first home robot that was affordable, smart and—best of all—useful. With the press of a button, the device vacuums a room and returns to its charging station when finished. It can detect and avoid walls, coffee table legs and stairs, and go where humans can't, like under the couch.

The Roomba became a hot Christmas gift and proved a hit for the company, which then expanded into other home robots such as the Scooba floor washer, the Looj gutter vacuum and the Verro swimming pool cleaner. On the backs of its dual military and consumer markets, iRobot turned a small profit in 2003 and has continued growing since. In 2005 the company went public on the NASDAQ stock market and in 2007, before the global recession soured virtually every industry, reported a profit of $8 million. The next year the company sold more than a million home robots worldwide, bringing its total Roomba sales to three million, and saw overall PackBot deployments reach 2,200.[7] The two markets have forced the company to learn different lessons, but in the end they are more similar than you'd think. "The consumer marketplace is very, very price sensitive. Everything needs to be engineered in an integrated fashion. There's no opportunity to put any fat into the design and still make money," Angle says. But both markets are utility-driven businesses. "If the Roomba doesn't actually clean your floor, we don't sell them. In the military, if the robot doesn't provide a tangible benefit to the soldier, such that the soldier is demanding to take the robot, then sales don't happen either. It's purely a utility sale as opposed to an entertainment or gadget sale because you don't buy vacuum cleaners on a whim, nor do you lug around a forty-pound robot for giggles."

One of the big differences in the company's two main products is their respective degree of autonomy. The Roomba operates with a high level of independence because the biggest problem that can arise is a tussle with the family dog. The PackBot and the company's other military robots, however, are remote controlled because troops cannot tolerate any unpredictability in their equipment like, say, having it turn on them *Terminator*-style. This is why autonomy for more sophisticated robots is coming in "on cat's feet," as Joe Dyer, the company's president of government and industrial robots and a retired navy vice-admiral, puts it. Dyer, a former pilot who displays the control stick from his old F-18 on the conference table in his office, explains that even today's super high-tech planes had all of their automation creep in one small piece at a time, starting with directional stability, then cruise control, then automatic landing. Autonomy for robots will happen the same way. "The next step of this is to say, 'Look robot, how about if you lose communications, you be bright enough to go back to where you can talk to us.' Autonomy really does come in on those cat's feet. It doesn't go to the governor of California."

The pace of that autonomy is likely to quicken over the next few years, since the current system of always keeping a man "in the loop" is tremendously inefficient. On a cost level, the benefits of replacing a human with a robot are voided if you still need that human to control the robot. On the consumer front, iRobot probably wouldn't have sold many Roombas if a human were still required to monitor the vacuum. Militarily, robots still have to transmit everything they see back to their operators, a considerable waste of wireless bandwidth, which is often at a premium in combat situations.

The solution to both issues is to give war robots more freedom by making them smarter. One possible scenario involves a sort of "swarm intelligence" where robots lead themselves. Imagine, for a minute, that a fleet of armed aerial drones is reconnoitering a hostile area when one of them is fired on by insurgents. Sensing that their cohort is under attack, the other drones head to its location to help out with the fight. Once the conflict is over, each individual drone resumes its original position, ready for the next encounter. Such independent swarm-minded robots may sound like the Borg from *Star Trek*, but they could act much more quickly and efficiently than they would if they were individually controlled by humans, who have to approve every move. Such systems, of course, will come in "on cat's feet" because of the inherent risks involved. "You've got to make sure you get it right because for a whole bunch of reasons you may not see it for another ten years if you screw it up," says Kevin Fahey, the executive responsible for purchasing robots for the U.S. army.[8]

Fighting the Future

It's developing this sort of autonomy, and doing so at a low cost, that was the whole point of DARPA's road races. In 2003 the agency announced the Grand Challenge, an ambitious 300-mile desert road race for unmanned cars between Los Angeles and Las Vegas. The contest was framed as a mini–Manhattan Project, with everyone from advertisers, corporate sponsors, science-fiction writers and even movie producers called on to get involved. "In order to make the DARPA Grand Challenge a success, we must maximize participation by everyone from major players to amateur enthusiasts," the program's manager said in announcing the race.[9]

The prize money was set purposely low at $1 million to encourage entrants to concentrate on cost efficiency—after all, no one was going to spend $10 million to win $1 million—while a desert course was chosen not only in case the machines turned on their human masters, but also because of the terrain's similarity to Iraq and Afghanistan. The inaugural race, held in March 2004, could have been considered a flop, since none of the fifteen participants actually finished—Carnegie Mellon's "Sandstorm" went the furthest, a whopping 7 of the 300 miles. But the entire idea was redeemed a year later. With the prize money raised to $2 million and the distance dropped to 130 miles, five of the twenty-three competitors finished the second race, led by Stanford University's "Stanley," which clocked top speeds of 40 miles an hour.

Six of eleven teams finished the third race, in 2007, led by the "Boss," despite the heightened complexity of driving in an urban environment, a city setting recreated on a closed military base in California. DARPA director Tony Tether was ecstatic about how much progress had been made in just three years. "The 2004 event was equivalent to the Wright brothers' flight at Kitty Hawk, where their airplane didn't fly very far but showed that flight was possible . . . The significant progress after 2004 was due to the fact that the community now believed that it could be done."[10] Sebastian Thrun, the Stanford artificial intelligence guru who led his team to victory in 2005, summed up the whole affair best: "We all won. The robotics community won."[11]

DARPA's main reason for holding the races was to help it meet a military transformation strategy mandated by the Pentagon. The process started in 1999, after the army experienced yet another embarrassing example of its own bloat and inflexibility. Confused logistics prevented the army from getting its Apache helicopters into Albania for use in the Kosovo War, which led to the conclusion that U.S. forces needed to get lighter and faster—and quickly. The Future Combat Systems plan, which featured a heavy reliance on small unmanned robotic vehicles, was drafted and approved by Congress in 2003. With an estimated budget of more than $200 billion and an expected completion date of 2030, military officials called it the most ambitious modernization of the army since the Second World War and the most expensive weapons program ever.[12] The plan called for the progressive introduction of unmanned air and ground vehicles, starting with a host of reconnaissance and explosives-disposal robots, then eventually moving to machines with weapons. With troops dying in Iraq and Afghanistan, it was easy to rationalize the inclusion of unmanned systems. "If you're at war, the Department of Defense asks for money, Congress moves it around. For the most part, people fund it. They want to support those in harm's way," Fahey says.

The rollout started slowly but ramped up fast. In 2004 American forces had 162 robots in Iraq and Afghanistan; by the end of 2008, they had more than six thousand.[13] At first, the robots came in cave-exploration and bomb-disposal forms (like iRobot's PackBot) and as aerial reconnaissance drones (like Northrop Grumman's *Global Hawk*). These were followed by the frighteningly named *Predator* and *Reaper* armed aircraft, built by General Atomics and operated by remote control thousands of miles away on a military base in Nevada. Armed ground robots such as Foster-Miller's MAARS and iRobot's Warrior have also been tested for battle. While the idea of having robots driving around with live guns scares some people, military experts say the evolution is a natural one that has happened with everything from cars to planes. "Almost every technology that finds itself in military service starts with reconnaissance and evolves to strike," says iRobot's Dyer. "It becomes so frustrating to be able to see but not to act, that it invariably moves to strike capability."

Future Combat Systems wasn't without its critics, however, including President Barack Obama, who inherited the program when he took office in 2009. Obama wasted no time in cutting back on FCS, and while his move was detrimental to big defense contractors like Boeing and Lockheed Martin, it is likely to pay off even more for small robot makers over the long run. Robert Gates, Obama's secretary of defense, announced in April 2009 that the government was cutting spending on "Cold War thinking"—areas of conventional warfare where the United States has a clear advantage—and would instead concentrate on new realities, like fighting the sort of urban-based counter-insurgents found in Iraq. Obama is "trying to take resources from areas where we have clear dominance—we control the skies and the seas—and move them to where we're challenged, which is irregular warfare and asymmetrical attack," Dyer says. "Robots are an important part of being able to meet that irregular warfare challenge. You can already see that with the IED threat. It is going to shift resources to an area that is advantageous to the robotics industry and iRobot in particular."

That could mean more programs like the DARPA road races. About 80 percent of the funding for artificial intelligence research in the United States already comes from the military, a percentage that could increase with the rethinking of defense spending.[14] One thing is for sure: the military's appetite for robots has only just been whetted and is very quickly becoming ravenous. As iRobot's Angle points out, the military market for robots is driving two relatively new concepts into the robotics industry: utility and cost effectiveness. While large Japanese car and electronics companies master the technology and have produced some amazing robots—such as Sony's AIBO robot dog or Honda's humanoid ASIMO—they have so far failed to produce robots that are cheap and useful. "The Japanese industry is a lot about 'cool.' AIBO and ASIMO and dozens of dynamically walking robot projects were developed so that large consumer electronics firms could show they can put together an impressive robot," Angle says. "Why is Matsushita cooler than Sony? Well, just look at their robots. It's bragging rights and it became a huge thing. The population got into it. It's misdirecting the Japanese into a world of show rather than a world of utility." Japanese officials don't disagree. "The U.S. is much better in the commercial field, but in technology, Japan is number one," says Takayuki Toriyama, executive director of the city of Osaka's office in Chicago. "The robotics market in Japan is not expanding right now because they're too expensive. Nobody can afford to buy a robot. This is a very serious problem."[15]

The United States is far from the only big military customer seeking cheap, useful robots. As of 2011, Unmanned Vehicle Systems International, an industry trade group, had 2,100 corporate members in 55 nations while a survey of government-related research found that at least 42 countries, including Britain, Russia, China, Pakistan and Iran, were working on military robotics.[16] The market for war robots is just starting to open, which means we are still at the early stages of seeing the commercial spinoffs—the Roomba was just the beginning.

A Smart Slut

While the overwhelming majority of artificial intelligence research is conducted on behalf of the military, some of it is coming from a surprising source: sex-minded criminals. Indeed, if military researchers don't come up with a human-seeming robot intelligence soon, hackers may very well beat them to it.

Internet scams started with spam—the unwanted email, not the meat in a can, that is. As with every other innovation on the Internet, the sex industry took quickly to it. Almost from the moment people started sending each other electronic messages, sex purveyors were acquiring email addresses to pitch their products. This resulted in the development of "spiders" that could trawl the web and search for email addresses and viruses that could infect inboxes and send out messages to all listed contacts. In the early days of the web, this sort of spam generally directed people to porn sites, where they would hopefully sign up for some sort of paid service. As anti-spam filters became smarter and stronger, the solicitations took ever-more complex forms. Simple spam morphed into annoying pop-up ads and then into phishing attacks, where a computer is infected by malicious code when the user clicks on a link.

In 2005, while I was living in New Zealand, I was nearly taken in by what was then the latest evolution of these scams. I had signed up to Friendster, a social-networking site that served as a precursor to the likes of MySpace and Facebook, and created a profile with all the standard information—my place of birth, age, interests and the like. While planning a visit home to Canada, I received a message from "Jen." She said she had read my profile and was interested in becoming a journalist and asked if I wanted to catch a Blue Jays baseball game when I was back in Toronto. Being a single guy at the time, I couldn't believe my luck—not only had someone actually read my profile, she also shared the same interests. I checked out Jen's profile and everything looked to be in order, so I replied and asked her to send more details about herself. She answered with a link to her website, saying there was information there. I followed the link and, sure enough, it was a site that required a paid membership to enter—clearly a well-disguised porn site. The jig was up.

After some Googling, I learned that many other men had been fooled by the same ruse. "Jen" had a different name every time, but "she" used the same script with individualized alterations gleaned from Friendster profiles. It turns out Jen was a sophisticated "bot" that was programmed to automatically scrub profiles for personal details, then try to pass itself off as a human in messages to users. Nobody ever did track down where that particular bot originated. I managed to trace the porn website as registered to a law firm in Australia, but my calls there were never returned. (Jen must have met someone else.)

The Friendster scam was small potatoes compared to the Slutbot, also known as "CyberLover," that made the rounds on dating websites in 2007. The "flirting robot" was a piece of software developed by Russian hackers that could establish relationships online with ten different people in just thirty minutes. The program, which could be configured into several versions ranging from "romantic lover" to "sexual predator," could carry on full conversations and convince people to reveal personal information by asking questions like, "Where can I send you a Valentine's Day card?" or "What's your date of birth? I'm planning a surprise for your birthday."

Security specialists said the artificial intelligence built into the software was good enough that victims had a tough time distinguishing the bot from a real suitor. "As a tool that can be used by hackers to conduct identity fraud, CyberLover demonstrates an unprecedented level of social engineering," one security analyst said. "Internet users today are generally aware of the dangers of opening suspicious attachments and visiting URLs, but CyberLover employs a new technique that is unheard of. That's what makes it particularly dangerous. It has been designed as a robot that lures victims automatically *without human intervention*"[17] (emphasis added).

The phenomenon provoked some bloggers to declare that the Turing test, devised by legendary British mathematician and computer scientist Alan Turing in 1950, had finally been beaten. Under Turing's test, a computer intelligence must fool a human judge—who is actively trying to determine whether it is in fact a machine—into believing it is a human. By some measures, CyberLover had beaten the test, but as other bloggers pointed out, "studies show that when people enter a state of sexual arousal their intelligence drops precipitously."[18] Moreover, many of CyberLover's victims were likely hoping it was a real person, and anyone in a state of sexual arousal is in no state to notice that the instigator of their excitement happens to be a computer.[19] While CyberLover may not have truly beaten the Turing test, it came close enough to provoke discussion.

Such online scams are certain to improve, especially since hacking has morphed in recent years from simple teenage mischief into a big profit-driven business. "They're really running it like a Fortune 100 company," says Dean Turner, global intelligence director for security giant Symantec. "Criminals are fundamentally lazy. They want to do the least amount of work to get the most financial gain. If that means they have to devote time and resources to working with groups who are working on things like artificial intelligence, they're probably going to do that."[20]

Porn You Can Feel

Disembodied artificial intelligences in the cyber-ether are great, but what happens when such programs can get up off the computer and walk around? The answer is obvious: sex robots. But before we get there, a few pieces of the hardware puzzle need to fall into place. The first piece comes from experimenters like Scott Coffman, who is what you might call a serial entrepreneur. The West Virginia native has in his lifetime drawn comics, published a board game, sold herbal supplements, invented paint-ball guns for kids and created the "Growl Towel," a small cloth waved by fans at Carolina Panthers hockey games. After dabbling in Internet porn for a few years, he finally found his calling when he launched the Adult Entertainment Broadcasting Network in 1999. The site took the innovative step of charging visitors for the porn they viewed on a per-minute basis, rather than the flat monthly fee most of its competitors charged. The approach worked amazingly well—AEBN has since grown into one of the busiest paid porn websites.[21] Coffman says AEBN has about 400,000 paid customers a month, brings in about $100 million in revenue a year and is the world's biggest video-on-demand company, in or outside of porn.[22]

It's no surprise, then, that Coffman believes entrepreneurship and innovation are the solutions to weathering the downturn in the porn business. To that end, he launched the

Real Touch sex toy in 2009 as his company's play against the encroachment of free online content. The Real Touch is about the size of a toaster, but it has the contours of a woman's hips, which comes in handy because the device is meant to be stuck and held on the penis. The inside is lined with warmed soft silicon that is moistened by a lubricant dispenser to simulate the feel of a vagina. Motors inside the Real Touch move in sync to specially coded movies, which can be viewed online when the device is connected to a computer. The $200 device is the latest in "teledildonics," sex toys that can interact with a computer, and Coffman is promoting it as the natural next step in porn. "Once you add the sense of touch to whatever that girl is doing to that guy in the movie, that's well worth paying for. That is the evolution of what adult entertainment should be."

The Real Touch uses something called haptic technology to introduce the sensation of touch to what is primarily a visual experience. It's not unlike the force feedback found in Xbox and Playstation controllers, or even in your cellphone when it's set to "vibrate." A haptic device uses sensors to detect when it is touching something, then relays that information back to its user, usually in the form of a subtle vibration, thus creating the sensation of touching something remotely. Like the porn industry, mainstream Hollywood is considering haptics as a way of creating an experience that can't be pirated. Montreal-based D-Box, for one, is now rolling out motion-synced chairs to movie theaters around the world. (The company also sells them for home use.) Watching a car chase, for example, takes on a whole new dimension as the D-Box chair rollicks and rolls in sync with the action on screen.

Non-entertainment concerns such as Quanser, another Canadian company based just north of Toronto, are also experimenting with haptics in fields such as surgery. Whereas traditional robotic arms allow operators to lift and manipulate objects, they are essentially lumbering oafs that don't convey any sense of "feel" and are ill-suited for precise or sensitive tasks. Quanser, which also builds unmanned aerial vehicles (UAVs) for the military, has designed an arm that uses haptic force-feedback technology in its fingers to relay the sense of touch back to its operator, who can "feel" what the arm feels using a bar-like control apparatus. I tried Quanser's hand at a robotics show in Boston, where I poked a surface with a pencil held in its grip, and it felt amazingly real. Haptic-enhanced robotics hold a world of promise, not only for surgery but also in the field of artificial limbs. Amputees have good reason to hope that they will soon be able to replace lost limbs with fully functional, "feeling" replicas. Such arms are indeed on the way, as we'll find out in the next chapter.

The other piece of the sex-robot hardware puzzle comes from companies such as California-based Abyss Creations, which began selling the Real Doll in 1996. While sex dolls in various forms have been used since at least the seventeenth century, the Real Doll reached a new level of technological sophistication. Using "Hollywood special effects technology" to create an amazingly lifelike female doll, complete with articulated steel skeleton and soft silicone outer layer for that "ultra flesh-like feel," the company is selling about 350 dolls a year at $6,000 a pop.[23] As of 2011 Abyss offered twenty different female models, some of which were indistinguishable from real women, at least in the photos. The company also offers three male dolls, named Michael, Nick, and Nate. All of the dolls, of course, have the necessary sexual openings—or appendages in the male version's case—and can be customized upon ordering. Those who can afford them have raved about their efficacy. "Best sex I ever had! I swear to God! This Real Doll feels better than a real woman!" exclaimed radio shock jock and sex aficionado Howard Stern.[24]

Abyss's success spawned a wave of imitators, both in the United States and beyond. In Japan, the sex doll business was already well established, largely because of a different cultural attitude toward such objects. While most Westerners think of sex with dolls as odd, creepy or pathetic, the phenomenon is considered far more normal in Japanese society. Japan has long accepted the standard reasons for having a sex doll—it can provide entertainment for men separated from their spouses for long periods of time and thus prevent marital infidelity, or act as an outlet for those who are unable to have sex with women for various reasons, such as physical issues or simple social ineptitude. Indeed, Japanese scientists achieved media fame in the sixties when they took inflatable sex dolls with them to the nation's research station in Antarctica. There are even, believe it or not, brothels in Japan where customers can pay to have sex *with dolls.*

The biggest sex doll maker, Orient Industry, sells about fifty a month, priced between $1,300 and $6,900, and exports them to Asia, Europe and the United States. Moreover, the company says the business is starting to go mainstream. "We aren't targeting 'otaku' or people with a doll fetish," a company manager says. "That boom has come and gone. Now we are getting a lot of healthy, normal people."[25]

"I Can Reprogram Her . . . to Like It"

It's only a matter of time before somebody puts all the existing pieces together. The realistic body of the Real Doll, the haptic feedback of the Real Touch, the artificial intelligence of the Slutbot, the humanoid movement abilities of Honda's ASIMO—add them all up and you could have a decent sex robot. An early sexbot wouldn't even need to be all that advanced to sell. "For guys, it doesn't have to have all the possibilities of true life, you're only looking for certain things. It doesn't have to be programmed to have tons of dialogue," says AEBN's Coffman. "I'm looking for the head rub, to be honest with you. If I could just have a robot that would rub my head all day after I got home, I'd be fine." In 2006 Henrik Christensen, chairman of the European Robotics Network, predicted humans would be having sex with robots within five years, or by 2011, which means we're just about there.

But are people ready for sexbots? After decades of fictionalized equivalents, the media is certainly ready for the real thing. When word got out in 2008 that Le Trung, a man living near Toronto, had built a lifelike female robot to help care for his aging parents, the worldwide media touted the story as "Man Builds Sex Slave Girlfriend." Trung, whom I found to be mild-mannered and polite when I interviewed him, was befuddled by the attention and feels wronged by how his creation, Aiko, which means "beloved one" in Japanese, was presented. "Tabloids need to make their money, right? The tabloids would ask, 'How long do you spend working on her?' and I said, 'Five hours a day.' Translation: I sleep with her five hours a day."[26]

Mind you, Trung didn't exactly help his cause by building sensors into Aiko's erogenous zones and programming her

278 Sex, Bombs, and Burgers

with the ability to simulate orgasms. When I met her, Aiko was programmed to bat my hand away if I tried to touch her breasts, but "I can reprogram her not to slap you, to like it," Trung said. Still, the robot's main purpose is to care and entertain. She can read medicine labels, announce the weather after checking it with her built-in wireless Internet connection, and even sing songs in Japanese. Impressive as she is though, especially considering she was constructed by one person with a budget of about $20,000, Aiko is quite frail. She can't walk or support her weight and her hands are made of cardboard stuffed into a pair of gloves. She could hardly have sex even if her creator wanted her to.

New Jersey software engineer Douglas Hines took it one step further with his True Companion, a robot that can talk and think and is anatomically functional. Hines thinks he's put most of the pieces together, save for the ability to walk (like Aiko), and started selling his creation in 2010 at a price comparable to the Real Doll. He says his robot, named Roxxxy, is more advanced than previous attempts because it can adopt different emotional states and personalities. "The sex side is easy, but nobody's integrated these pools of technology," Hines says. "Roxxxy takes the inputs she's given and decides what emotional states they're associated with. If there are enough inputs given to justify a transition to that state, then she transitions. So, for example, if she's sleeping and hears that you're waking up and trying to interact with her, she'll make that transition if she has enough input over enough time to the sleepy state, and then continue on from there."[28]

Whether the True Companion proves to be a hit or not, most people will still consider sex with robots to be the same as sex with dolls—either odd, creepy or pathetic. But sexual psychologists argue that this attitude is bound to change. Consider that only sixty years ago, homosexuality, pre-marital sex and masturbation were all generally considered wrong and immoral. Now, early into the new millennium, all three are more or less accepted. Even the staunchest conservative American states are allowing gay marriage, no one bats an eye when an unmarried couple moves in together, and vibrator sales are going through the roof. Even more recently, online dating was viewed with considerable disdain, as a refuge for the desperate or socially maladjusted, but no one speaks ill of it anymore because, well, everybody is doing it.

In his 2007 book *Love + Sex with Robots*, British artificial intelligence researcher David Levy argues that having sex with and even marrying robots will be commonplace by 2050, for both men and women. Others, including inventor and futurist Raymond Kurzweil, believe it will be even earlier, perhaps by 2029. Levy says we will fall in love with our robots for the same reasons we fall in love with other humans, our pets or even inanimate objects like cars or computers. Like *Star Trek*'s Commander Data, they can be programmed to be just as intelligent, funny, romantic and caring as any human. Levy argues that robots will be even better sex partners than humans because they can be programmed with all the sexual information in the world. Imagine an encyclopedic knowledge of the *Kama Sutra*.

Robots could also provide "spice" to a relationship by learning their mate's behavior and varying their programming accordingly, perhaps by changing their voice, personality or even appearance. And they'll be able to fulfill sexual fantasies in ways that real people cannot. As one example, many couples jokingly grant each other a "get-out-of-jail-free" card, a permission to cheat on their partner with the celebrity of their choice should the unlikely opportunity ever present itself. While a man in such an arrangement may never get to have sex with the real Angelina Jolie, he certainly could with a reasonable facsimile. That may seem far-fetched, but Abyss is already making Real Dolls designed in the likenesses of real porn stars working for adult company Wicked Pictures. Licensing one's image to sex robot makers is a potentially huge source of revenue for celebrities and porn stars alike.

When sex robots do arrive and attitudes toward them change, one thing is definite: they *will* sell. If the historical market for prostitution and pornography is any indication, robots may very well end up being the best thing to happen to sex since the discovery of the orgasm. Some even suggest that the old fear about robots, that they will steal jobs from humans, will come true—in prostitution. "When sexual robots are available in large numbers, a cold wind is likely to blow through the profession, causing serious unemployment," says Levy.

Somehow, I don't see Toyota going anywhere near this.

Replacing Teenagers

Teenagers may also be on the endangered list, as least as far as employment is concerned. The boring, repetitive and unskilled jobs that are the hallmarks of many people's adolescence are ripe picking for robots, too. R. Craig Coulter knows this, which is why he's helping fast-food companies adopt robotic labor.

Coulter, a PhD graduate of Carnegie Mellon University's prestigious robotics program, knows exactly what his advanced university degree amounts to: $6.50 an hour. It could have been worse, though. That wage was actually twenty-five cents more than the typical employee got at McDonald's, which is where he worked after completing his robotics degree. "I had dinner with the dean of the school of computer science at Carnegie Mellon a year after that happened, and I told him I had empirical evidence for what a PhD is actually worth," he laughs.[29]

All kidding aside, Coulter didn't work at McDonald's because he couldn't find a job after university. He was conducting research on HyperActive Bob, a robotic order-taking system that he hoped would transform the fast-food industry. The idea for Bob came to him after seeing one too many orders screwed up by staff at fast-food drive-thrus. Like many who have experienced this, Coulter couldn't could help but think, "How hard can it be to get an order right?" The problem, he found while working his McJob, was that despite being one of the world's biggest industries, the fast-food business still depends far too much on low-paid, low-skilled human labor. The employees, usually teenagers, are often uninterested in the repetitive and dreary work. While every other industry has actively investigated how to put robots into such jobs—which the military refers to as the "3 Ds," for dull, dirty and dangerous—the fast-food business has been uncharacteristically slow to look at new technology. "It's the last $100-billion-a-year industry on the planet that hasn't automated," Coulter says.

With that in mind, Coulter founded HyperActive Technologies with his friend Kieran Fitzpatrick, a fellow Carnegie Mellon graduate. In 2001 the duo consulted with fast-food industry analysts and found that not much could be done to improve the efficiency of the actual restaurant kitchens. They also concluded, however, that visitors became customers as soon as they entered the restaurant's property—in many cases, the parking lot—yet nobody engaged with them until they reached the counter to place their order. The space in between was valuable time that could be spent preparing food for the customer.

The first version of Bob tried to address this problem with a set of sensors on the restaurant's roof that detected new vehicles as they entered the property. A software program then tried to anticipate what each vehicle's occupants might order. A minivan, for example, probably meant children were on board, so kids' meals should be prepared. This sort of vehicle profiling proved too inaccurate, though, so Coulter and Fitzpatrick focused solely on visitor volumes instead. When fed with enough historical sales data, the new system could accurately predict what menu items would be needed within the next few minutes and beam the information to employees inside the restaurant via an interactive touch-screen. "Statistically, it would be very difficult with an individual coming into the restaurant to say, 'That guy is going to want a cheeseburger,'" Coulter says. "But if you've got ten people coming into the restaurant, you know that two or three of them are going to want a cheeseburger and some of them are going to want chicken."

HyperActive tested the system with McDonald's, Burger King, Taco Bell and a few other fast-food giants before finding an anchor customer in Zaxby's, a mid-sized American chicken chain, in 2005. Zaxby's prided itself on serving customers only freshly cooked chicken, which often resulted in long wait times or wasted food. HyperActive's system, which costs about $5,000 to set up, reduced wait times by giving staff a better approximation of how much food would be needed, and when. The chicken chain estimates that stores using Bob save about $5,000 a year in food waste, and they also see other, less-tangible benefits, such as fewer employees quitting. "The turnover is relatively high because people don't like getting yelled at. When Bob goes in, the yelling goes away," one restaurant owner said.[30]

The big chains haven't adopted Bob because HyperActive, as a small company, doesn't yet have the ability to supply the systems in the volumes needed. With Zaxby's, HyperActive has been able to roll out Bob on a restaurant-by-restaurant basis. McDonald's, for its part, seems to resent the suggestion that it even needs a robot system. A manager for the chain's Canadian operations told me his restaurants were opting instead for a staff-scheduling system based on sales statistics. "In a lot of cases, that's generally too late for you to react to it, which is why we're doing more on the proactive side of the measurement and the projections of what our sales are going to be."[31]

The fast-food giant's suppliers and other food processors are considerably less resistant to using robotics. Lopez Foods and Tyson Foods, two of McDonald's suppliers of beef and chicken, respectively, each use robotics—one to package and stack burger patties, the other to refrigerate poultry without human intervention. The market for such industrial robots, which have long been almost the exclusive domain of carmakers, is growing quickly as manufacturers add in new capabilities. Iceland-based food-processing equipment maker Marel, for example, has perfected a robotic system that can wash, de-slime, de-head, skin and filet about twenty fish a minute, while Germany's Carnitech has built a fully automated boat that can process and pack 500 tons of shrimp a month. Industrial robot maker KUKA, also based in Germany, has helped a meat processor replace its manual butchery with a fully automated system that uses lasers to track carcass sizes and positions. Industrial food-processing robots made up only 3 percent of the total market in 2009, but the share is growing as processors slowly come to appreciate the military's "3 D" mantra.[32]

Independent food companies are also starting to get into the robotic fast-food game. Toronto-based Maven's Kosher Foods has designed a vending machine that can dispense freshly cooked hot dogs, while in Italy an entrepreneur has rolled out the strangely named "Let's Pizza," a similar contraption that creates custom-made pizzas in three minutes for five euros a pop. A glass window on the machine lets the buyer watch as the pizza dough is mixed and spun into shape, the sauce and toppings are added and the pizza is then cooked to taste. An American company, La Pizza Presto, has the same idea, but its machine cooks pre-made pies in just ninety seconds. When asked for their thoughts on the new inventions, Italian pizza cooks were predictably dismissive. "You can't make any comparison, especially in terms of quality. The only benefit is the price," one told Reuters.[33]

They couldn't be more wrong. While chefs like to consider food creation an art inseparable from human touch, robots are actually perfectly suited to the task. They can't come up with their own recipes—yet—but they can certainly replicate ones they are programmed with, and do so perfectly, every time. Humans, who are prone to distraction and errors, can't make the same claim.

The big fast-food chains can't afford to ignore robotic technology indefinitely because, as several have learned in recent years, the risks of continuing to rely on low-paid, low-motivated workers may prove too great. In 2008 Burger King was hit with a public relations fiasco when an employee took a bubble bath in one of the restaurant's large sinks, then posted a video of it on his MySpace page. Similarly, Domino's faced public outrage in 2009 after a video of employees putting snot into food hit YouTube. Both companies had to backpedal and assure customers that these were isolated incidents and that their food was indeed clean and safe. The damage to their reputations, however, was done.

As more of these incidents surface, and they inevitably will given the ubiquity of social media such as YouTube, the appeal of replacing low-paid labor with robots is only going to increase. That means teenagers and other unskilled laborers will be displaced and have to find a new kind of work. It may be a small price to pay to keep boogers off our pizzas.

OPERATION DESERT LAB

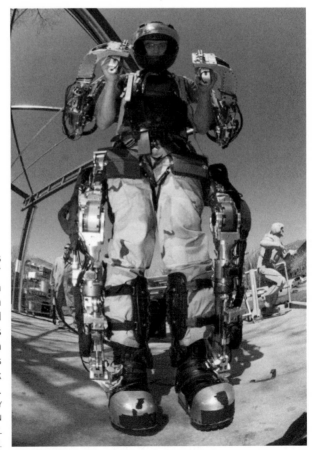

Raytheon's "Iron Man" exoskeleton can lift a hundred kilos and is agile enough to allow its wearer to kick a soccer ball.

For much of modern history, the Middle East has been a fertile land of invention where science and technology flourished. From the seventh century onward, while Europe wallowed in the war, hunger and disease of the Middle Ages, the Middle East enjoyed a golden era, an enlightened renaissance from which a steady stream of life-improving inventions flowed. Water turbines, navigational astrolabes, glass mirrors, clocks, the fountain pen and even an analogue computer that calculated the date were just some of the innovations of the time. While Europeans were busying themselves burning libraries and fighting over who God favored, Islamic scholars were laying the foundations for many of the world's modern institutions by opening the first hospitals, pharmacies and universities. They also laid the groundwork for a number of modern sciences, including physics, chemistry, mathematics, astronomy and medicine.

Things have changed dramatically in recent times. The past few decades of political turmoil, war and, in some places, religious fundamentalism have largely crippled the region's intellectual institutions. An area that was once the envy of the world for its progressive thinking now lags in just about every intellectual and technological measure. In Iran, the former seat of the once-powerful Persian Empire, literacy rates are well below that of the Western world. In Iraq and Afghanistan, the current centers of conflict, barely 40 percent of the population can read. Internet use is well behind the developed world, and where people are actually surfing the web, censorship is rampant. While the blocking of pornographic sites isn't too surprising in Muslim countries, the definition of "questionable" content also extends to political and free-speech websites and tools. During Iran's election turmoil in 2009, the popular messaging service Twitter was blocked to prevent details of uprisings from spreading. Egypt completely disconnected itself from the Internet in early 2011 when similar demonstrations took place. Spending on science and technology in the region stands at a woeful 17 percent of the global average, ranking not just behind the West, but also behind some of the poorest countries in Africa and Asia.[2]

Technological advances have occurred in the Middle East in recent years, but perversely, they've been deployed by Western militaries. Since the early nineties, the United States and its allies have used the area as a sort of a laboratory for a vast array of new technologies, testing out their capabilities to see what works, what doesn't work and what can be improved. The impetus behind all new war technology, the military tells us, is to save lives, but as we've already seen, there's also the important by-product of technological spinoff into the mainstream, which is a key driver of Western economies. The recent conflicts in the Middle East are perhaps the best example to date of this terrible duality of military technology: while new war tools and weapons inflict tremendous pain, suffering and hardship on one group of people, they also create prosperity, convenience and comfort for others.

Another Green Revolution

Iraq's invasion of Kuwait in 1990 sparked a new wave of Western technological development. The ensuing liberation through Operation Desert Storm was of course motivated by oil interests, but it also provided an opportunity for the American military to field-test new technologies, some of which had been sitting on the shelf from as far back as the Vietnam War.

Smart bombs, or precision-guided munitions, were the natural opposite of dumb bombs which, when dropped from a plane, simply used gravity to find their target. Smart bombs were developed during the Vietnam War and used lasers to find their mark—the target was illuminated by a beam that the bomb homed in on. The new weapons promised two key advantages over their precursors: they could improve the efficiency of bombing missions by decreasing the number of munitions needed, thus saving on costs and maximizing damage, and they could lower so-called "collateral damage," or the destruction of non-military targets and civilian deaths. American forces used such bombs in small numbers in Vietnam, as did the British military during the Falklands War of 1982, but they proved to be of limited use in poor weather. It wasn't until the Gulf War that they were improved and deployed on a large scale.

General Norman Schwarzkopf, the American commander of the coalition forces, set the tone of the war in January 1991 when he dazzled reporters with a videotape of a smart bomb zooming through the doors of an Iraqi bunker to blow up a multi-storey command center. The early stages of the war were going exactly as expected, Schwarzkopf announced, thanks largely to the incredibly accurate bombs being used. "We probably have a more accurate picture of what's going on . . . than I have ever had before in the early hours of a battle," said the veteran general, who began his military service way back in 1956.[3] Like radar back in the Second World War, smart bombs were hailed by an impressed media as a "miracle weapon" that pounded the Iraqi military into a sorry state, making the ensuing ground war short and easy. Only 7 percent of the munitions dropped on Iraqi forces, however, were of the "smart" kind; the rest were the traditional variety.

Still, smart bombs had proven their worth and their usage has steadily increased in each subsequent conflict. Fully 90 percent of the bombs brought to Iraq by American forces for the second go-round in 2003 were "smart."[4] The laser guidance used in the weapons, meanwhile, has gone mainstream in recent years, primarily in cars, where it has been incorporated into collision-avoidance systems. Toyota, for one, introduced a laser cruise control system in its 2001 Lexus which, like many of the robot vehicles in the DARPA road races, used beams of light to track other cars ahead of it.

If the videos of smart bombs flying into Iraqi targets in stunning first-person view were not enough to impress the public, most of whom were watching it all unfold on CNN, then the images of night-time air attacks were. The scenes I remember best involved the volleys of Iraqi anti-aircraft fire arcing upward at unseen stealth bombers high above. First, an orchestral cascade of lights would fly into the sky, followed shortly thereafter by a brilliant, expanding explosion on the ground. It seemed to be clear evidence of which side was winning. Like many people watching, I was awed by the technology and not thinking of the lives lost. All of it, of course, was broadcast in its full green-tinged glory.

Night vision was another technology that had been sitting around for a while. The earliest version of it was invented by the American army during the Second World War and saw small-scale use in sniper-rifle scopes in the Pacific. About 300 rifles were equipped with the large scopes, but the poor range of only about a hundred yards limited them to defending the perimeter of bases. Nazi scientists also developed night-vision "Vampir" rifles and mounted similar units on a few tanks. The problem with both versions was that they used large infrared searchlights to illuminate targets so that gunners equipped with scopes could see them. This gave away the searchlights' position, making them easy targets.

By the Vietnam War, American scientists had improved the technology to use available light, such as moonlight, which again limited use to when weather conditions were good. By 1990 the technology had entered its third generation and evolved to use "forward-looking infrared" (FLIR) image intensifiers, which electronically captured and amplified ambient light onto a display, such as a television monitor or goggles.

A FLIR device displays a monochrome image, usually green or grey, because it uses light from just below the spectrum visible to the eye. The technology therefore needs no additional light sources and functions well in any sort of weather. The new goggles were small, light, low-power and cheap (you can buy them online today for a couple hundred dollars), which is why the U.S. Army bought them by the truckload for Desert Storm. Night-vision was also incorporated into a lot of the military's sensor and video technology, including the cameras that captured those green-tinged bombing images broadcast on CNN. If smart bombs were miraculous, the night-vision sights used by pilots and the goggles worn by ground troops were even more so, because they allowed coalition forces to "own the night." "Our night-vision capability provided the single greatest mismatch of the war," said one American general.[5]

After the war, night-vision technology was adopted quickly by the mainstream, particularly in security. Parking enforcement, highway rest stops, tunnel surveillance, transit systems, ports, prisons, hospitals, power plants and even pest inspectors all found it amazingly useful. The spread of night vision closely paralleled the rise of digital cameras, which also underwent their baptism of fire during the Gulf War. Both technologies became remarkably cheap, remarkably fast and began to converge, with night vision becoming a standard feature of video cameras early in the new millennium. As prices continued to drop on both technologies, they soon became standard in just about every camera available, which means that anyone can now create their own green-tinged Paris Hilton–style sex video.

On the military front, night-vision technology continues to evolve, with scientists currently working on doubling the field of view and adding thermal-imaging abilities to goggles. Lord only knows what sort of sex videos will come out of that.

The "Technology War"

The coalition forces had one other fancy new enemy-locating technology at their disposal: the Global Positioning System or GPS we have all grown to know and love. GPS units allowed troops to locate enemy positions and movements with pinpoint accuracy, further increasing the efficacy of smart bombs and units equipped with night-vision. It was a new holy triumvirate of American weaponry that reinforced the old saying, "You can run, but you can't hide." If one technology didn't find you, the others would.

GPS had its origins in Navsat, a satellite navigation system first tested by the U.S. Navy in 1960. The original system used five satellites and only provided a fix on the user's location once an hour. The technology was slowly upgraded throughout the seventies and early eighties, when tragedy hit. In 1983 a Korean Air Lines flight was shot down for straying too far into Soviet air space, prompting President Reagan to declare that the GPS, which could prevent such disasters, would become available for civilian use when it was completed. The twenty-four second-generation GPS satellites were scheduled for launch between 1989 and 2000. When the war started in 1991, however, only sixteen had been launched, eight short of the required number to provide worldwide coverage. Nevertheless, the incomplete system—which still provided three-dimensional navigation for twenty hours a day—was pressed into service at Schriever Air Force Base in Colorado. In the Kuwaiti desert, which was largely devoid of landmarks or way posts, GPS finally delivered on space technology's long-held promise of making conflicts on Earth easier to fight. "It was the first war in which space systems really played a major role in terms of the average soldier, sailor, airman and Marine," said a general with the U.S. Air Force Space Command. "This was the first time that space affected the way our troops fought in the battle."[6]

With the war concluded, President Bill Clinton signed the system's dual military-civilian use into law in 1996. Civilian access, however, was not as accurate as the pinpoint precision enjoyed by the military, a discrepancy the government fixed in 2000 to enhance GPS's usefulness to the public. The amendment made consumer GPS devices ten times more accurate, to the point where location could be determined within a few yards. "Emergency teams responding to a cry for help can now determine what side of the highway they must respond to, thereby saving precious minutes," Clinton said. "This increase in accuracy will allow new GPS applications to emerge and continue to enhance the lives of people around the world."[7]

Right there to take advantage of the newly opened market was Garmin, a company started in Kansas in 1989 by two electrical engineers. Gary Burrell, a native of Wichita, and Taiwan-born Min Kao—the company's name is a contraction of their first names—spent much of their early careers working for military contractors. Their first product, launched in 1990, was a dashboard-mounted GPS for marine use that sold for $2,500, while a follow-up handheld unit proved popular with troops in Desert Storm. The company wasted no time in jumping into the consumer market and has since ridden the quickly growing market to riches. As of 2008 the company had sold more than forty-eight million personal navigation devices,[8] more than half of the total worldwide market, which is expected to continue growing by 20 percent a year until 2013, when it will exceed $75 billion.[9] A substantial part of that growth will come from the current wave of smartphones, which since 2004 have incorporated cellular-assisted GPS chips.

All of the new sensor and navigational technology meant there was a ton of electronic data pouring in, but this could only be turned into useful intelligence if there was some way to crunch it all. As luck would have it, the war coincided perfectly with the rise of personal computers. While the first desktop computers were made available in the late seventies, sales didn't

296 Sex, Bombs, and Burgers

begin to really ramp until the early nineties when the devices became standardized and simple enough for the average user. Among the first major business or "enterprise" buyers was the U.S. military, which used PCs during the Gulf War for everything from organizing the movement of troops to sorting through satellite photos for targets. American brass even used computers to simulate Iraqi responses to their battle plans, some of which turned out to be more efficient than the actual reactions.

By 1990, the U.S. military was spending $30 billion a year on desktop computers.[10] One of the biggest beneficiaries of this huge outlay was Microsoft, which was in the process of standardizing the operating system that personal computers run on. The third version of Windows, released in 1990, was the first to have a slick visual interface that required minimal training to learn, and the first to really take off with users. By 1991 Windows 3.0 had sold more than fifteen million copies, a good portion to the military, and helped give Microsoft three-quarters of the operating system business worldwide.[11] After combat ended, General Schwarzkopf gave appropriate credit. Calling Desert Storm "the technology war," he said, "I couldn't have done it all without the computers."[12]

The Gulf War was indeed the first technology war and it set an interesting precedent. The entire conflict lasted less than three months, while the ground campaign took only a hundred hours. Coalition casualties were low and the end result was total victory for American-led forces. It was easy to view the Gulf War as the perfect war, if such a thing could exist. Certainly Schwarzkopf saw it that way and his crediting of new weapons and tools as the key to victory cemented a long-held belief in American policy, that technology was the country's biggest advantage over the rest of the world—not just in war, but in business as well.

The average Joe sitting on his couch watching the war on CNN couldn't help but agree. The Vietnam War had dragged on for years, took a high casualty toll and looked like plain hell, with its grime and misery. The Gulf War, on the other hand, was quick, painless, and quite simply, looked good on TV. The green-hued battles, the first-person bomb views, the alien-looking stealth bombers—it was like a slick, Michael Bay–directed science-fiction movie come to life. Americans came home every night from work and turned on the tube to watch their boys, decked out in all the latest high-tech gear, kick Iraqi butt. It was a far cry from seeing them flee in disgrace from Vietnam. No wonder American morale, both military and public, was riding high after the war. Technology had re-established itself as America's not-so-secret weapon.

Humans 2.0

American morale, however, took a massive blow with the destruction of the World Trade Center in the terrorist attacks on September 11, 2001. Whatever smugness was left over from the Gulf War quickly turned to a desire for revenge. The people responsible for the attacks, the al Qaeda terrorist network, would feel the full fury of America's technological arsenal. Once again, the weapons developed in the decade since the Gulf War would be unleashed for field testing against the country's enemies.

While the weapons were first deployed in 2001 in Afghanistan, where al Qaeda's leader Osama bin Laden was apparently hiding out, they again found their way to Iraq with a second invasion in 2003. The situation again played itself out as if it were a movie, a sequel to the original Gulf War. And, like all sequels, the second go-round was bigger, louder, deadlier and ultimately less successful than the original. The new two-front war posed some unique challenges to the American-led coalition forces. In Afghanistan, the enemies were holed up in mountains and caves, making it difficult to find them and get to them. In Iraq, the invasion was relatively quick, but it soon morphed into an urban insurgency where guerrillas disguised themselves as civilians and melted into the city throngs. Smart bombs, night vision and GPS were handy, but new-and-improved technologies were needed to tackle these challenges.

Some of the technologies deployed so far in Afghanistan and Iraq, like the robots we heard about in chapter nine, have already begun to spin off into commercial uses. As both conflicts drag on, however, more and more new technologies will be developed and tried out. With some of them, their potential commercial applications can only be guessed at, while others have more obvious uses. One example of the latter are bionics and prosthetics, which are in some ways by-products of all the work being done on robotics. DARPA's work on prosthetic limbs to help amputee veterans has been nothing short of spectacular. Scientists working for the agency have in recent years designed a *Terminator*-like robotic arm that displays a range of function far beyond that of even the best conventional prosthetics, allowing the user to open doors, eat soup, reach above their head and open a bottle with an opener, not to mention fire and field strip an M16 rifle. The prototype arm only weighs about eight pounds, has ten powered degrees of freedom with individually moving fingers, eleven hours of battery life, and like the arm I tried out at the robotics conference in Boston, has force-feedback so the amputee can actually "feel" what he or she is touching. The arm entered advanced clinical trials in 2009 with commercial availability expected to follow soon after. The next phase of the program, DARPA says, is to implant a chip in the patient's brain so that he or she can control the arm neurally. The chip will transmit signals to the arm wirelessly, allowing the amputee to manipulate the arm with mere thoughts. The idea isn't science fiction—the FDA is fast-tracking tests, which means the arm could be commercially available by 2015.

In a similar vein, defense contractor Lockheed Martin unveiled its Human Universal Load Carrier exoskeleton system in 2009. The HULC, which is the only name it could possibly have, allows soldiers to carry up to 220 pounds with minimal effort. Presumably, like the green-skinned comic-book character, it will also let them "smash puny humans." The exoskeleton transfers weight to the ground through battery-powered titanium legs while an on-board computer makes sure the whole thing moves in sync with the wearer's body. The HULC is surprisingly nimble, too, allowing the soldier to squat or crawl with ease.

The exoskeleton is designed to alleviate soldier fatigue while carrying heavy loads across long distances, but Lockheed is already investigating industrial and medical applications, some of which are obvious. Like the giant exoskeleton used by Sigourney Weaver in *Aliens*, a juiced-up version of the HULC could easily find work in docks and factories where heavy lifting is required. Honda, in fact, is testing a similar system, but its device has no arms, only legs. The carmaker's assisted-walking device, which looks like a bicycle seat connected to a pair of ostrich legs, is designed to support body weight, reduce stress on the knees and help people get up steps and stay in crouching positions. For workers who spend the whole day on their feet, like hairdressers, such a device would be fantastic. "This should be as easy to use as a bicycle," a Honda engineer said. "It reduces stress and you should feel less tired."[13]

Not to be outdone, Lockheed competitor Raytheon—the folks who brought us the microwave—is also getting in on the action. The company says it too has an exoskeleton that, like Lockheed's, can lift hundreds of pounds but is also agile enough to kick a soccer ball, punch a speed bag and climb stairs with ease. The company began work on the system in 2000 when it realized that "if humans could work alongside robots, they should also be able to work inside robots."[14] The media dubbed Raytheon's exoskeleton "Iron Man," after the comic book character, which should prove the perfect foil to Lockheed's HULC.

Scientists working for the military have also made significant strides in health and medicine over the past few years. DARPA researchers have even managed to come up with a simple cold medicine. In trying to alleviate the cold and flu symptoms soldiers often experience after strenuous exertion, researchers discovered the natural anti-oxidant Quercetin. In an experiment that involved three days of hard exercise, they found that half of one control group became ill with colds and flu. The incidence in the other control group, which took the anti-oxidant, was only 5 percent. Quercetin has since been commercialized as RealFX Q-Plus chewable pills.

On a grander scale, DARPA is also influencing the way pandemics are fought. The impetus behind the agency's Accelerated Manufacturing of Pharmaceuticals program was to greatly reduce the length of time between when a pathogen is identified and when a treatment is widely available, which has typically been a very long fifteen years. DARPA is seeking to reduce that to a mere sixteen weeks or less, simply by changing the way vaccines are produced. While treatments have traditionally been grown in chicken eggs, DARPA scientists are experimenting with growing them in plant cells and have found that a single hydroponic rack, about sixteen feet by ten feet by ten feet, can produce sufficient protein for one million vaccine doses, thus doing the work of about three million chicken eggs at a fraction of the cost. Moreover, the plants are ready within six weeks of seeding and produce vaccines that eggs can't, like one to fight a strain of avian flu. The new technique is inspiring pharmaceutical companies to try different approaches. In November 2009 Switzerland's Novartis, for example, won German regulatory approval for Celtura, an H1N1 vaccine manufactured using dog kidney cells.

There's also a gizmo known simply as "the Glove." It looks like a coffee pot except it has a cool-to-the-touch metal hemisphere inside, where users place their palm. Researchers at Stanford started working on the device in the late nineties and got DARPA funding in 2003. They had developed the theory that human muscles don't get tired because they use up all their sugars, but rather because they get too hot. When users place their hand inside the Glove, their body temperature cools rapidly, allowing them to resume in short order whatever physical activity they were performing. The net result is that the user can exercise more. "It's like giving a Honda the radiator of a Mack truck," says Craig Heller, the biologist behind the device.

One of Heller's lab technicians incorporated the Glove into his workout regime. When he started, he was managing 100 pull-ups per session, but by using the device he was able to do more sets. Within six weeks, he was doing 180 pull-ups and in another six weeks he was doing more than 600. Heller himself used the Glove to do 1,000 push-ups on his sixtieth birthday.[15] The Glove's military uses are obvious—because it effectively duplicates the effects of steroids, which allow users to train harder and more frequently, it's going to result in stronger and faster soldiers. Its commercial applications are also apparent; every athlete and gym in the world is going to want one. If we thought athletes jacked up on steroids were playing havoc with sports records, wait till they get hold of the Glove. The device also has humanitarian potential, because it works in reverse. It can rapidly increase body temperature as well, which means it could save people suffering from hypothermia and exposure.

Then there's the questionable stuff. DARPA has historically steered clear of biological research, but everything changed after September 11. Tony Tether, DARPA's director at the time, adopted a more open attitude toward bioengineering and picked a fellow named Michael Goldblatt to lead the charge. The move was perhaps the best example yet of a bombs-meets-burgers crossover as Goldblatt had spent more than a decade working for McDonald's, most recently as the company's vice-president of science and technology. The same man who tested low-fat burgers for McDonald's was all of a sudden in charge of bioengineering better soldiers. Goldblatt even referenced his past while spelling out his priorities at the annual DARPAtech convention in 2002:

Imagine soldiers having no physical limitations . . . What if, instead of acting on thoughts, we had thoughts that could act? Indeed, imagine if soldiers could communicate by thought alone or communications so secure there is zero probability of intercept. Imagine the threat of biological attack being inconsequential and contemplate for a moment a world in which learning is as easy as eating, and the replacement of damaged body parts as convenient as a fast-food drive-thru.[16]

The idea of bioengineered soldiers has been around for decades, mostly in the realm of science fiction. Marvel Comics' Captain America, the product of an injected "super-soldier" serum, and the Hollywood stinker *Universal Soldier* starring Jean Claude Van Damme and Dolph Lundgren as genetically and cybernetically jacked-up commandos come to mind. For much of the past decade, some of this science fiction has become, as Goldblatt calls it, "science action." DARPA is funding dozens of "human augmentation" projects around the world, all of which are geared toward changing the old army slogan "Be all you can be" to "Be more than you can be."

Scientists at Columbia University in New York, for example, are working on using magnetic brain stimulation to lessen a person's need for sleep. Defense contractor Honeywell is using electroencephalographs to detect neural spikes in satellite analysts' brains before they consciously register what they are seeing, which is resulting in faster action. Boeing is using near-infrared technology to monitor pilots' brains with the hopes of eventually allowing them to fly several planes at once. At the University of Alabama, scientists have successfully kept lab mice alive with 60 percent of their blood gone through injections of estrogen, a process they believe can be replicated on humans.

And that's just the stuff we know about. While many of the publicly known programs stop short of full-out genetic engineering, there's little reason to believe that some cloning research *isn't* being done with military applications in mind. Congress has raised some concerns over this sort of biological research, but the net effect so far has been the delay of funding for some projects or the simple renaming of others. "Metabolic Dominance," for example, was changed to the less ominous-sounding "Peak Soldier Performance."[17]

Found in Translation

Like any good business, the military is looking to streamline operations, cut costs and introduce efficiencies. That's the impetus for the U.S. military's "network-centric operations," or a fully networked battle force. The plan is to get all those robots in the field to communicate better with their human masters, and also with each other. It also means coming up with better ways to deal with the ever-increasing amounts of data pouring in from the battlefield. Part of the solution is better communication systems, such as the delay-tolerant network Vint Cerf is working on. Another idea is cognitive computing, an effort to reduce the amount of data humans have to sift through by letting computers do it for them. The computer then only passes on the most pertinent details to its human master, perhaps with a suggested course of action. This rudimentary artificial intelligence is intended to reduce the military's "tooth-to-tail" ratio—the number of support staff it has to field compared to its actual fighting forces.

In 2009 Robert Leheny, DARPA's acting director, made the case for smarter computers in a speech to a House of Representatives committee on terrorism. "Without learning through experience or instruction, our systems will remain manpower-intensive and prone to repeat mistakes and their performance will not improve," he said. The Department of Defense "needs computer systems that can behave like experienced executive assistants, while retaining their ability to process data like today's computational machines."

DARPA's Personalized Assistant That Learns program, or PAL, is doing just that in military hospitals. The computer system is capable of crunching large amounts of data and then taking action by itself. Receptionists, not programmers, are teaching the system to find vacant appointment slots and make referrals.

If that sounds like the beginning of a *Terminator*-style apocalypse, the work being done on translation is where things really get scary. Franz-Josef Och looks like your average mild-mannered computer programmer, although his German accent might frighten some into believing he is the quintessential mad scientist. He grew up in a small town near Nuremburg and discovered a passion for computer science early on. Around 1997, while attending the University of Erlangen-Nuremberg, Och became interested in something called statistical machine translation, a method of understanding languages using algorithms rather than grammatical rules.

The idea of using computers to translate languages has been around since the beginning of the Cold War, when the United States was focused on understanding Russian and vice versa, but very little quality progress was made over the intervening fifty years. The problem, Och explains, was twofold. The grammatical approach, in which a computer is programmed with the rules of two languages, say English and Russian, doesn't work very well because there are too many little differences, slang uses and idiosyncrasies to provide an accurate translation. The statistical approach, in which a computer algorithm analyzes patterns in the languages and then compares them, was potentially more promising, but it also had big issues. First, Cold War–era computers didn't have the processing power to analyze reams of digitized data. Second, those reams of data, which are all-important in producing the statistical sample for algorithms to analyze, simply didn't exist. By the late nineties, however, both problems were no longer issues as computer processors were packing impressive horsepower and the Internet had made digital data plentiful.

In 2002 Och went to work at the University of Southern California's Information Sciences Institute. The same year DARPA sponsored a contest, not unlike its robot car races, to develop new algorithms for statistical machine translation, particularly ones that could translate Arabic into English. Och entered the contest in 2003 and built a translation engine using publicly available documents from the United Nations, which are automatically translated by actual humans into the six official languages: Arabic, Chinese, English, French, Russian and Spanish. With this gold mine of millions of comparable digitized documents, Och's algorithm scored an impressive accuracy rate and won the DARPA prize. The following year he was snagged by search-engine giant Google which, like the military, had a significant interest in computerized translation. The two actually have reverse interests: while the military wants to translate languages, particularly Arabic and Chinese, into English so that it can understand its real and potential enemies, Google wants to translate English into other languages to open up the largely Anglo web to the rest of the world, which will dramatically increase the size of its advertising market.

Google launched Translate in 2001 and refined it with Och's methods when he came on board. As of 2011 this online tool, where the user simply pastes in the foreign text and presses a button to have it translated into the language of their choice, handled more than fifty languages. And it works—unlike those other gibberish-spouting attempts that came before it, Google Translate generally gives you the gist, and a little bit more, of the text. As a result, the search company tends to score at the top of annual machine-translation tests run by the National Institute of Standards in Technology. Still, the tool isn't perfect and the trick now is to get its success rate up to 100 percent. To that end, the company in 2009 announced "translation with a human touch," in which actual human translators can suggest improvements to the algorithm's results. The success rate is also bound to improve, Och says, now that the tools to create better systems are freely available. "It's so easy now because some PhD student somewhere can download all the data and some open-source tools that various people have written and build an end-to-end system. It was virtually impossible before."[18]

Google is also applying the statistical approach to voice translation. In 2007 the company launched a 411 phone service that people could call to find businesses they were looking for. The caller spoke their query into the phone and Google computers would either speak the answer back or send a text message with the information. The point behind the service was to amass a database of voice samples, similar to Och's U.N. documents, from which Google's algorithm could work. The experiment has borne fruit, with the company launching a voice search service for mobile phones that lets users speak their query into the device, rather than typing it.

DARPA is now testing similar technology in Iraq. Soldiers there are being equipped with iPod-sized universal translator machines that can interpret and speak Arabic for them. The devices have the basics down, like discussions about infrastructure or insurgents, and are slowly improving in other areas as well. "We knew that we couldn't build something that would work 99 percent of the time, or even 90 percent of the time," says Mari Maeda, who runs the translation program for DARPA. "But if we really focused on certain military use cases, then it might be useful just working 80 percent of the time, especially if they don't have an interpreter and they're really desperate for any kind of communication."[19]

Some linguists believe that computers, which have already become better chess players than humans, will eventually surpass our ability to translate languages as well. "Human translators aren't actually that great. When humans try to figure out how to translate one thing, they drop their attention as to what's coming in the next 'graph," says Alex Waibel of Carnegie Mellon, who was also born in Germany and does translation work for DARPA. "And they're human. They get tired, they get bored."[20]

Welcome Our Robot Overlords

The potential for machine translation far outstrips simple military and commercial uses. The ability to understand all languages may take the Internet's equalizing and empowering abilities to a higher level, providing a greater chance at world peace than we've ever known. Since the advent of mass media, the public has had its opinions of other people in other parts of the world shaped largely by third parties: newspapers, books, radio, television, movies. While the Internet opened up direct links between the peoples of the world and theoretically cut out those middlemen, the language barrier still prevents real communication. With truly accurate and instantaneous text and voice translation only a matter of years away, that final obstacle is about to fall. In a few years time Americans, for example, will no longer have to take the media's portrayals of Middle Eastern Muslims at face value. They'll be able to read, watch and understand Arabic news as easily as they view the *New York Times'* website or CNN. And people from around the world will be able to communicate and interact with each other directly, one on one. Pretty soon, we'll be getting friend requests on Facebook and Twitter from people in China, Tanzania and Brazil. Our social circles are about to broaden massively and we're going to learn a lot more about people who have been alien to us thus far. As Och puts it, "There's a real possibility to affect people's lives and allow them to get information they otherwise couldn't get. Machine translation can be a real game-changer there. That seems to me to be a good thing."

Indeed, with this greater communication will come a greater understanding of other people, which will make it more difficult to go to war against them. If governments—the democratically elected kind, anyway—find it difficult today to muster public support to attack another country, it will only be harder when there are direct communications between the people of those two countries.

Where things really get interesting is in the application of statistical machine translation to more than just languages. Because the algorithm is designed to identify patterns, its potential uses in artificial intelligence are mind blowing. Google has identified as much and is taking baby steps toward the idea. In 2009 the company announced plans for a computer vision program that will allow machines to identify visual patterns. The project, still in its research phase, uses the same sort of statistical analysis as Translate. Google fed a computer more than forty million GPS-tagged images from its online picture services Picasa and Panoramio and came up with a system that could identify more than 50,000 landmarks with 80 percent accuracy. In announcing the project the company said, "Science-fiction books and movies have long imagined that computers will someday be able to see and interpret the world. At Google, we think computer vision has tremendous potential benefits for consumers, which is why we're dedicated to research in this area."[21]

Science fiction is in fact proposing the next direction that statistical machine translation could take. *Caprica*, the short-lived prequel to the hit series *Battlestar Galactica*—the best show ever, I might add—explored the idea of using pattern-identifying algorithms to create an artificially intelligent (AI) personality. In *Caprica*'s pilot episode, a teenager named Zoe uses such an algorithm to create a virtual AI of herself by feeding it with all the personal digital data she has produced in her lifetime. After Zoe's death her father, roboticist Daniel Graystone, discovers the AI in a virtual world created by his daughter. The AI, a perfect replica of Zoe, explains to him how it was done:

AI Zoe: You can't download a personality, there's no way to translate the data. But the information being held in our heads is available in other databases. People leave more than footprints as they travel through life. Medical scans, DNA profiles, psych evaluations, school records, emails, recording, video, audio, CAT scans, genetic typings, synaptic records, security cameras, test results, shopping records, talent shows, ball games, traffic tickets, restaurant bills, phone records, music lists, movie tickets, TV shows, even prescriptions for birth control.

Daniel: A person is much more than usable data. You might be a good imitation, a very good imitation, but you're still an imitation, a copy.

AI Zoe: I don't feel like a copy.

As we learned with sex robots in the previous chapter, the lines between real, thinking and feeling human beings and well-programmed machines are likely to blur in the future. If, as in Zoe's case, a computer can be programmed to statistically infer how individuals might act based on everything they've done before, we may very well be forced to treat them as real people. Och, who has watched *Battlestar Galactica*, isn't sure that his algorithms will eventually result in killer Cylon robots, which is what Zoe's AI eventually becomes, but he does think they will enable smarter machines. "Many people see different things in the term 'artificial intelligence,'" he says, "but it will definitely lead to more intelligent software."

The Future Is Invisible

While many of the technologies outlined in this chapter are already having an impact on the world outside the military, there are also some way-out-there lines of research that could lead us in directions we've never even dreamed of. In that respect, Sir John Pendry, a theoretical physicist at London's Imperial College, may one day be viewed as the "godfather of invisibility."

Pendry is an institution in British physics, having received numerous honors over his decades of work in optics, lenses and refraction, culminating in his knighting by the Queen in 2004. In 2006, he explained to the BBC his theory of how invisibility would work:

> Water behaves a little differently to light. If you put a pencil in water that's moving, the water naturally flows around the pencil. When it gets to the other side, the water closes up. Special materials could make light 'flow' around an object like water. A little way downstream, you'd never know that you'd put a pencil in the water—it's flowing smoothly again. Light doesn't do that of course, it hits the pencil and scatters. So you want to put a coating around the pencil that allows light to flow around it like water, in a nice, curved way.[22]

It turns out that's not so hard to do after all. The key is something called a metamaterial, a composite material constructed at a macroscopic level rather than at a chemical level, which is how a traditional substance such as copper is made. Because they are designed on a much smaller level, metamaterials can conduct electromagnetism in ways not found in nature. Not surprisingly, DARPA has taken an interest in metamaterials and in 2004 held a conference in Texas to discuss potential uses of these new substances. Pendry was invited to give a presentation, wherein he suggested that metamaterials could be used to influence electromagnetic forces to bend light. He wrote the idea up for DARPA—the agency occasionally sponsors research from non-American nationals, as long as they are friendly to the United States—and since then, two separate groups at Berkeley and Cornell universities have used the idea to build metamaterial "invisibility cloaks." The cloaks were only a few millimeters wide and could cover only two-dimensional objects, but they successfully bent light to flow as a fluid around their subjects. With proper funding, which will doubtlessly come from the military, it will only be a few years until large, stationary three-dimensional objects can be made invisible. Moving objects, however, are more complex, so it may be a while before Harry Potter's fabled invisible cloak becomes a reality. The first realization will probably be a static object, more like Harry Potter's gun turret than Harry Potter's cloak.

Metamaterials offer a world of possibilities. Bending light to confer invisibility may be just one of their nature-defying capabilities. More applications will become apparent only as scientists come to understand them better, as will the mainstream benefits. One spinoff is already being seen. Because they are tremendously light, metamaterials are working their way into radar systems, making these less bulky and extending their use to new applications such as the car collision-detectors mentioned earlier. As iRobot's Joe Dyer says, these sorts of far-out technologies tend to come in "on cat's feet," one small step at a time.

That said, when it comes to invisibility, the British military isn't waiting for the technology to creep in slowly. In 2007 the U.K. Ministry of Defence and its contractor QinetiQ announced it had successfully made tanks invisible, albeit not with metamaterials but with more pedestrian technology. Using video cameras and projectors mounted on the tank itself, QinetiQ researchers fooled onlookers into completely missing the vehicle by projecting its surroundings onto its surface, an advanced form of camouflage. "This technology is absolutely incredible. If I hadn't been present I wouldn't have believed it," said a soldier present at the test. "I looked across the fields and just saw grass and trees, but in reality I was staring down the barrel of a tank gun."[23] British military experts expect such tanks to be in the field by 2012—which, from the way things are going, means they could see action in Afghanistan or Iraq.

And yes, it is delightfully ironic that British scientists, who worked so hard to make things visible with radar sixty years ago, are now putting so much effort into making them disappear from view.

The Pornography of War

It's a paradox that the longer the wars in Afghanistan and Iraq go on, the more technological advances there will be. In a way, the more death and destruction the West visits on the Middle East, the more economic benefits it will reap, since the weapons of today are the microwave ovens and robot cars of tomorrow. Like hunger and poverty, the desire to test out new technologies is a major driving force behind such conflicts. The more politically minded would-be terrorists see this unfortunate cause and effect as a form of imperialism, so they fight it by joining al Qaeda and the Taliban. For the general population, it must also rankle. Rather than creating the inventions, as they've done for centuries, the people are instead having foreign technologies tested on them. It's a far cry from the Islamic Golden Age of science and reason.

Such is the way of war, however. When two opponents are evenly matched, there is less likelihood of one side trying anything radical; while a crazy new technology may promise ultimate victory, it can also bring disaster. Far-out new technology is usually only deployed after it has been thoroughly tested, or when one side has an apparent advantage. The Second World War is a perfect example. Nazi Germany only started deploying its futuristic weapons, like the experimental and highly volatile V-2 rockets and jet fighters, when the tide of the war had turned against it, while the United States only dropped the atomic bomb once victory was a foregone conclusion. The longer a war goes on, however, the less apparent it is that one side or another has an advantage, which is certainly the case in Afghanistan and Iraq.

The United States government, however, still believes it has the edge in those conflicts, so it continues the steady rollout of new technologies. For the immediate future, American defense spending will focus on smaller, more flexible and more personal technologies. Small, light robots will be a priority and individual soldiers will get a lot of new equipment to help them find terrorists who have melted into the urban landscape. Some of this technology will be biological, like the DARPA experiments into areas such as regeneration and heightened cognition, while some of it will be oriented around communications and sensors.

Since the first conflict with Iraq in the early nineties, we in the West have come to believe that technology is a key factor in deciding who wins a war. Images from the Second World War, Korea and Vietnam painted horrific pictures: soldiers suffering in disease-ridden trenches or military hospitals with their limbs blown off, bodies being carried off the battlefield, dirt-smeared faces. Recent conflicts, however, have all but erased those images and replaced them with scenes of laser-guided bombs, futuristic-looking planes, bloodless and victimless destruction, green-tinged battles of lights. War has become sanitized, safer, almost fun. While troops in the Second World War had to sing to each other to raise morale, soldiers in Iraq can now while away their leisure time playing Xbox games. To those of us who are insulated from the day-to-day horror, war is more like a game— or at least it is sold as such, if the Air Force's video-game-heavy website is any indicator.

This sanitization and video-game-ification has affected the general public. With his smart bomb video, Norman Schwarzkopf gave rise to the "war porn" phenomenon, which has grown in lockstep with the rise of the web. YouTube is rife with videos depicting American tanks, jets and drones blowing up Iraqi and Taliban targets, many set to rollicking heavy-metal soundtracks. The same is true in reverse, with the website full of videos of insurgents setting off explosives and blowing up American forces. While only a small subset of the population enjoys watching such footage on a regular basis, the reaction of many to a video of a *Reaper* destroying a building tends to be, "Oh, neat." Never mind the people inside that building who have just been obliterated.

This is how we subconsciously want to deal with war. We know it's happening and we know the real human cost, but we prefer to think of it as a necessity that produces "neat" results. That has certainly been the case in the Middle Eastern desert, where the death of thousands of innocents over the past twenty years has indirectly supplied us with more comfort and convenience than we know.

The Benevolence of Vice

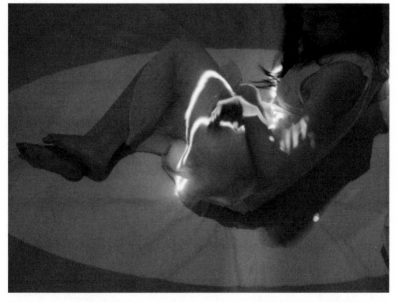

The Mutsugoto system is designed to allow people to have sex with one another remotely by "drawing" lights on the user via an Internet connection.

There's a scene in the movie *Fight Club* where the camera pans around Ed Norton's apartment and labels pop up on screen to describe each piece of furniture, indicating what it is and how much it costs. That's kind of what the world has been like for me during the course of writing this book, only I'm not seeing an IKEA catalog come to life, but rather the connections to war, sex and food. I see them everywhere I turn, or I end up suspecting connections and looking them up. I'm surrounded by them in my home—a very typical home, I like to think, which means that you're probably surrounded by them too.

In the living room, there's the plasma television, which saw its origins as a computer display. In their early days, computers were the exclusive domain of the military. My TV's manufacturer, Panasonic, known as Matsushita in Japan, was shut down briefly after the Second World War when the Allies realized it had profited throughout the conflict by building everything from radios to bicycles for the Japanese government. The DVD/CD player sitting under the TV, plus all of my discs that play in it, are based on lasers, the first of which was created by Hughes Research Laboratories, a defense contractor, back in the late fifties. Lasers are everywhere now, but like computers they were once primarily meant for military use. There's also the fact that the whole home entertainment market got a healthy kick-start from porn. Back when Hollywood was busy suing VCR makers, porn producers were pumping out videotapes. It's not too crazy to suggest that if they hadn't done so, the home video market, and therefore DVD players, may never have developed.

Next to the DVD player sit perhaps my two favorite pieces of electronics, the Xbox 360 and the Playstation 3. For both, we can thank Ralph Baer and his defense contractor employer, Sanders Associates. Every time I sit on the couch and blow people up online while playing *Call of Duty*, I can't help but think of the irony. I don't think Vint Cerf, when he was helping to launch the ARPAnet, imagined that's what people would eventually be using it for. Then again, maybe he did.

My couch is stuffed with memory foam, a substance designed by NASA in the sixties to improve aircraft cushions. The mattress in the bedroom also contains it. The air conditioner in the window owes a good deal of its history to Willis Carrier, an American inventor who perfected his cooling methods during the Second World War when he came up with a system that could simulate the freezing, high-altitude temperatures found on military planes. After the war, he used the techniques to help launch a boom in residential air conditioners. I'm glad he did because the summer months can get really sticky.

In my office, just about everything is derived from war and sex. The computer is perhaps the best example. Apple co-founders Steve Jobs and Steve Wozniak got their start as summer interns at Hewlett-Packard, which by the seventies was a big supplier of computers and electronics to the military. In fact, the duo left and started Apple because HP didn't want to get into consumer products. That's ironic now, given that HP is the biggest personal computer manufacturer in the world. Of course, what's a computer without an Internet connection? As we've seen, that's a technology that was military made and porn perfected. Heck, even the mouse was created with DARPA funding.

Over on the wall hangs a whiteboard on which I scribble my thoughts and to-do lists. The main ingredient in it is melamine, a plastic that, like many plastics, was first put to wide use during the Second World War as dishware aboard navy ships. Recently, melamine acquired a bad rap because some Chinese food makers sneaked it into baby formula. The plastic shows up in chemical tests as a protein, so it's the perfect filler for unethical food makers looking to cut corners.

Just about everything in the bathroom is made by large conglomerates that also have significant food operations. The Dove soap and Q-tips are from Unilever, which counts Ben & Jerry's ice cream, Lipton drinks and Ragu spaghetti sauce among its many food brands. The Pantene shampoo, Crest toothpaste, Gillette razors and shaving gel, Oral-B toothbrush and floss are all from Procter & Gamble, which also makes Pringles potato chips and Folgers coffee. It's a little discomforting to know that the same companies that are formulating my shampoo are also concocting what I put in my belly. I'm also a little wary of using my Right Guard deodorant now that I know where it comes from. The brand is owned by Henkel, which, like many German companies, used concentration-camp prisoners as labor during the Second World War. The product name is a little off-putting knowing that.

The kitchen, meanwhile, may as well be a lab. There is, of course, the microwave, supplemented by the ancient Westinghouse stove and refrigerator. Westinghouse is a company rich in military history, having designed radar systems, jet engines and scores of other war technologies. All the cleaners and detergents under the sink are made by chemical companies that became wealthy supplying the military. Even the humble aerosol can, which houses air fresheners, spray paints and other fun substances, was derived from that most fearful of war weapons, the flame thrower. The food itself is a cornucopia of technology, from the frozen fries and strawberries in the fridge to the retort pouches of soup on the shelves, not to mention the chemically-laden Little Debbie's chocolate Swiss rolls I love so much. Even the fresh apples have a thin veneer of wax to make them look shinier, while the bananas and tomatoes have been artificially ripened with ethylene gas—which the fruits produce themselves, albeit in smaller amounts, anyway.

That's my home, in a nutshell. I suspect it's pretty typical, a real-world example of the comforts that have evolved from the technology of humanity's vices. It's proof that technology is neutral—it's what we do with it that matters.

The Good That Bombs Do

The atomic bomb, the most destructive invention in history, is a good example of technology's dual nature. The bombs dropped on Japan during the Second World War killed hundreds of thousands and instilled a chill that caused millions to live in fear for decades. But the technology behind the weapon has evolved over the years to the point where it now promises to heal as well as harm.

Lawrence Livermore National Laboratory, about fifty miles east of San Francisco, has been researching nuclear weapons for more than half a century. The lab was founded in 1952, in the midst of the Cold War, to augment the work being done at the Los Alamos base in New Mexico, where the original bomb was created. Physicist Ernest Lawrence, a key figure in the building of the weapon, had set up a lab at the Berkeley campus of the University of California modeled on the process used in the Manhattan Project. His multidisciplinary team of scientists pursued large-scale research, leading to some early successes including the creation of small nuclear warheads that could be launched from submarines, more powerful computers and strides in fusion energy.

In the sixties Livermore scientists broadened their research into peaceful uses of nuclear power and the effects of radiation on humans, and in the seventies they expanded into lasers. Today, the lab is one of the pre-eminent laser facilities in the world, prompting some to joke that LLNL doesn't actually stand for Lawrence Livermore National Laboratory but rather Lasers, Lasers 'n' Lasers. In 1992, after the United States stopped nuclear testing, Livermore, Los Alamos and the Sandia lab in New Mexico were charged with overseeing the Stockpile Stewardship Program, an ongoing project to maintain the safety and reliability of the country's nuclear arsenal.

Livermore has produced a plethora of weapons technology. But, as with all such science, there has been a considerable upside as well—the lab has churned out beneficial mainstream technologies by the truckload. Dyna, a collision-modeling software program, is just one example. Developed in 1976, the program gave scientists a way to measure how bombs respond to ground impacts, or how the weapons' nose cones react when they hit. This sort of data is vital in munitions design because engineers need to know how the weapons will explode against different surfaces such as concrete, sand or metal. Early versions of the software did simple numerical analyses, but as the eighties came and computers acquired graphical interfaces, these evolved into visual representations. It didn't take Livermore scientists long to realize the possible commercial applications of the software, and in 1980 they began sharing it with industry.

Car companies were the first to jump on board. Because actual car-crash simulations are expensive—up to $1 million or more each—the companies were keen to try out the virtual equivalent. "They realized that they can do practical crash analysis using a simulation tool," says Ed Zywicz, one of Livermore's Dyna programmers.[2] As the software's code is available for free, it spread quickly through the industry and has been in perpetual development since, with improvements shuttling back and forth between the lab and companies. So are carmakers contributing to building weapons by giving code improvements back to Livermore? "Definitely," Zywicz says. "There's give and go in both directions."

The software, and variations such as the French-designed Radios, is now being used by just about every company that needs to predict collisions. It is used by train makers, by Boeing to see how bird impacts affect jet engines and even by Coors to simulate beer-can mishaps on production lines. It's hard to gauge how many lives Dyna has saved, but it must be many.

Livermore's biggest contribution to the civilian sector has been in genetics, a field it got into naturally when it began exploring the effects of radiation on human physiology. In the early nineties, scientists at the lab made a huge breakthrough with the invention of "chromosome painting," a process that allows researchers to label individual chromosomes with fluorescent "tags," thereby making them more visible. The technology, which was licensed and made available to industry relatively cheaply—one "painting kit" for twenty tests sold for just $400 in 1992—greatly simplified the identification of disease and genetic defects. Through the nineties, scientists used kits to identify the genes behind a host of health issues, including muscular dystrophy, kidney disease, migraines and dwarfism, which helped in the development of many treatments.

Livermore's genetic research was a key factor in launching the Human Genome Project, an initiative to map the full human DNA sequence. Scientists at the lab began mapping out chromosome nineteen, one of the twenty-three pairs humans have, in 1987. When the worldwide project was launched in 1990, Livermore was given the responsibility of mapping out three full chromosome pairs. The project was completed in 2000, ahead of schedule, and the results are only now starting to be understood. This research has deepened our fundamental understanding of human biology, and its potential applications are vast. In the years ahead, scientists may use the knowledge that stems from it to stamp out many diseases, with the ethical debates about whether they should do so sure to follow.

Today, Livermore sprawls over a 1.3-square-mile campus and employs more than 7,000 people, half of whom are scientists, with an annual budget of about $1.5 billion. Researchers there are now promising to revolutionize treatment of that age-old scourge of humanity: cancer. With all sorts of lasers and other beams at their disposal, they have developed a cancer treatment method that uses protons instead of x-rays. The technology was first used to determine the safety of nuclear missiles by scanning their insides with proton beams, but researchers eventually discovered the same trick could be applied to humans.

The problem with x-rays, according to George Caporaso, who leads the lab's proton project, is that they aren't very accurate when blasting cancer cells. If you hit someone with an x-ray beam, a lot of the rays are absorbed by the surface of the skin while others penetrate beyond the cancer cells, deeper into the body than desired. Proton beams can be modulated so that they enter the body at a low frequency, spike in intensity just as they hit the cancer cells, then quickly wane afterward. The difference, which is "revolutionary, not evolutionary," is akin to using a scalpel rather than a baseball bat to slice away cancer cells, Caporaso says.[3] "I'm not a medical physicist, but from what I know about it, it is paradigm-changing."

In 2007 Livermore commercialized the technology through a licensing agreement with TomoTherapy, a Wisconsin-based cancer treatment company. The proton-accelerator machines are still huge and expensive—they're as big as a basketball court and cost upward of $200 million, which means that as of 2009, there were fewer than thirty treatment centers worldwide and therapy was generally only available to the wealthy. Caporaso and his team are working to bring the size and price tag down over the next few years. They're aiming for a machine that is only six feet long and costs $20 million, which will make proton cancer therapy available to the general public.

There probably isn't a better example of technology's dual nature than Livermore. On one hand, scientists there have contributed inventions that can destroy the world many times over; on the other, they're working diligently to save lives and improve the world. Zywicz sums it up best: "It's amazing how all the defense applications over the years have spun off into things like Dyna, which are so vital and useful to the civilian section. Without that defense research spending, it never would have happened."

Enemies Today, Friends Tomorrow

War-inspired technology isn't just helping with health problems, it's also paradoxically presenting new opportunities for peace. While some believe a superpower showdown between the United States and China is inevitable, technologists think such a war is highly unlikely because of how interwoven the two countries' economies have become through Internet-based supply chains. If the United States went to war with China, the flow of manufactured goods would stop dead. Americans would be forced to do without everything, from new computers to clothes to diapers. China, meanwhile, would see its biggest market close its doors. Both countries would plunge into an economic meltdown and take the rest of the world with them, making the Great Depression look like a gentle dip in the stock market.

When the world was less interdependent, wars were commonly started over a desire for more territory, as was the case with the Second World War. That seems silly now. If technology-driven economies in tiny countries such as Luxembourg and Ireland can pull in more gross domestic product per capita than big nations such as the United States, Canada or Australia, land grabs simply don't make sense anymore. Resource grabs, such as the American invasion of Iraq, are still happening, but technology may provide the answer there too. If the developed world can wean itself off oil—hybrid and electric cars are the early steps in this process—peace may finally come to the Middle East. We are starting to move beyond wars fought by nation against nation and toward those fought by nations against marginalized groups. That seems to be a step forward. If only we could solve the growing problem of food shortages.

And let's not forget the final barrier of language is set to fall, thanks to technology. The people of the world are about to become a lot closer. The Internet united and inextricably linked the economies of China and India with Western countries; we're about to take the next step to deepening such relationships, which means there is hope that the future will bring better understanding, and perhaps fewer wars.

A Porn Star President?

Pornography has had its negative effects, not the least of which was paving the way for talentless hotel heiresses to achieve fame. But advances in the technology used to sell porn have also brought us closer together. As a trail-blazing communications medium, albeit an often naughty and sometimes obscene one, pornography has funded the development of technologies that let us connect with one another. Whether it's laying phone lines in developing countries, giving VCR makers a market or adding functionality to the Internet, porn companies have often been willing to put down the dollars when no one else would, and we've all benefited from their actions. And there's no reason to believe we won't continue to do so. Here's one example. Researchers at Scotland's Distance Lab, an organization devoted to better connecting people separated by long distances, are working on something called Mutsugoto, an Internet-based system that lets users have sex with each other through remote-controlled lights. On one end, the user sets up an electronic light rig over his or her bed, which then connects through the Internet to another system anywhere else in the world. Users on both ends wear special rings that control the lights over the other's bed, which they then use to sensually "draw" on the other person's body. As the Distance Lab puts it, "drawings are transmitted 'live' between the two beds, enabling a different kind of synchronous communication that leverages the emotional quality of physical gesture."[4]

It's a way-out-there idea, but it's also a fascinating take on communications technology that only hints at untold-of possibilities. Like AEBN's Real Touch and other teledildonics, applications of sexual technologies are poised to take communications in directions we've never considered.

With new connections opened up, we are in fact communicating better, particularly on sexual issues. As pornography has gradually encroached on mainstream media, we have been forced to re-evaluate and re-define what is socially permissible every step of the way. While some people lament how sexualized our society has become today, bringing sex out to a place where it is openly discussed has surely been one of the most positive developments of the past century. Gay rights, though they still have a long way to go, simply didn't exist a few decades ago, while being a single or divorced parent used to earn social scorn. Though porn and technology are by no stretch of the imagination responsible for all of this, they have played key roles. As porn star Joanna Angel says: "I don't know if I can give porn the credit for really starting the chain, but it's definitely part of the chain."[5]

Porn technology is continuing to change our views on sex. As Paris Hilton and *Girls Gone Wild* have shown, porn is expanding from its traditional role as a professionally produced product that is sold to consumers to something that anyone can partake in. Social websites such as MySpace, ubiquitous cameras and cellphones are allowing for the instantaneous mass spread of amateur porn and giving young people a very different attitude toward sex and nudity from the one their parents likely had. Stoya, the young porn star from Philadelphia, says she got into the business professionally because half the people she knew were putting naked photos and videos of themselves on the Internet just for fun anyway. "It's not a big deal. The kids I meet don't necessarily look at it as these big scary adult companies holding guns to people's heads, exploiting them and forcing them to do things they don't want to do, because they have a friend or have talked to someone on MySpace [who's involved]."[6]

Porn is also becoming today's rebellion of choice. While a few decades ago a teenager might have bought a leather jacket or smoked cigarettes to rebel against her parents, today she just might make her very own green-tinged sex video and post it online. You never know—tomorrow's president may not be asked whether he inhaled marijuana, but whether he wore a condom.

Are these positive developments? Some argue that the sexualization of culture is a sure sign of decadence and decline. Many more think that sexual liberation and acceptance of different lifestyles are positive steps forward in our evolution.

A Paradox of Plenty[7]

Of our shameful trinity of desires, technology has produced the most dramatic transformation of opinion in the domain of food.

In the fifties, food was much scarcer than it is today and wasn't available year-round. As a result, meat sat frozen in the freezer for months while vegetables and other preserves waited on the shelf in jars and cans. People would typically eat just about anything they could get their hands on.

Half a century later, abundance has turned us into a very selective society. Some consider us complainers. While people in starving countries will do just about anything for a steak or a pork chop, many in the West are turning to vegetarianism and veganism. As comedian Chris Rock so bluntly joked, "We got so much food in America we're allergic to food. Allergic to food! Hungry people ain't allergic to shit. You think anyone in Rwanda's got a fucking lactose intolerance?"[8] We're even choosier about what we eat than who we have sex with. As one sociologist puts it, "To compare junk food to junk sex is to realize that they have become virtually interchangeable vices— even if many people who do not put 'sex' in the category of vice will readily do so with food."[9]

What is often overlooked is that technology provides us with plentiful food that is inexpensive, available year-round, easy to store and fast to prepare and throw away if we don't want it or use it in time. This has formed the backbone of all the other freedoms we enjoy. The success of the McDonald's restaurant in the heart of Manhattan's financial district is proof—at lunchtime, it's packed with the high-powered financiers and stock brokers who keep the wheels of the global economy spinning (although not so well, of late). Big Macs and Quarter Pounders provide them with the quick calories they need to get on with their busy day. Ironically, food technology has created a luxury of riches that gives us the *choice* of being vegetarians and vegans.

Despite the issue of obesity and the related health problems it can bring on, abundant food, along with improved medicine, has also significantly increased how long we can expect to live. In 2011 the average lifespan of Americans, Canadians, Brits and Australians, the people who eat the most processed foods, was between seventy-eight and eighty-one. That represents an increase of about ten years from the fifties, when mass processing really kicked in. Advances in medicine helped, but maybe all those Big Macs and Twinkies haven't been so bad for us after all.

We want our food to be healthier, with fewer preservatives and chemicals, but we don't want to give up the convenience or low costs we've come to expect. As such, we're inspiring even more technology. Food scientists are obliging with new advances like the Natick lab's pressurized processing. Scientists at the University of Alberta, meanwhile, are experimenting with replacing chemical preservatives with natural ingredients such as mango pits and the fatty acids found in wheat and barley. These natural substances are turning out to be just as good at destroying harmful bacteria as their chemical counterparts. "If you replace chemicals with a natural preservative, without compromising safety, the [food] quality is better," says one scientist.[10]

Beyond the ever-shifting appetites of people in the developed world flutters the specter of a growing population and shrinking farmland. Developing nations are also, well, developing—their food systems are catching up to those of the West. The amount of technology that goes into our food is only set to grow.

If You Can't Beat 'Em, Join 'Em

War, porn and fast food don't exist in a vacuum. It would be wrong to say these three businesses have driven *all* technological advances—many other industries have contributed their fair share, and will continue to do so.

Computer and software giants such as Microsoft and Apple have developed scores of innovations, from the Windows operating system to the ubiquitous iPod. Soon we'll be touching our computers and talking to them instead of typing and mousing. Carmakers have given us steadily better products, from power steering to collision detection, and they've created some remarkable robots. In the years ahead, our cars will drive themselves while we surf the Internet on a dashboard-mounted, voice-activated computer. Pharmaceutical companies have developed a raft of miracles, with drugs that treat everything from blood pressure to erectile dysfunction. In the future, we'll have better cold medicines and faster vaccines for potential pandemics. Telecommunications firms have connected us in myriad ways, from the Internet to cellphones. Mobile chips inserted in our heads to give us an instant Internet connection cannot be too far off. Hollywood studios have given us home video and are now introducing three-dimensional and motion-synced movies. Entertainment is becoming more immersive, and we are edging closer to fulfilling the early promise of virtual reality.

These industries are just as competitive as their military, porn and fast-food counterparts. They also depend on innovation to drive new products and thereby profit growth. But for the most part, we consider them clean industries, inspired by entrepreneurs, ingenuity or generally noble needs. We don't look down on them the way we do on our shameful trinity. But what's funny is that none of them really is clean. As should be apparent by now, virtually every industry has directly or indirectly benefited from the innovations of war, sex and fast food.

In the end, we're no closer to creating a society bereft of our shameful trinity of needs than we were a thousand years ago. There's every reason to believe that the technologies of war, sex and fast food will continue to shape our world. Over the past century, these industries have been primarily America-driven, but that's changing as the world becomes increasingly globalized.

Food is already a global industry and will only become more so as more countries develop their markets. Demand for porn is global but there are very few international players like Playboy, largely because the industry is dominated by relatively small, privately run companies. The Internet-driven decline in the business has everybody talking about consolidation, so many smaller players may have to merge into fewer bigger companies to remain competitive. As for war, you'd think it would be the one market that would never go truly global because of national security concerns, but it's already surprisingly open. Historically, NATO and the Warsaw Pact shared military industries, but today the market is realigning itself from West versus East to nation states versus terrorists. This reshaping may produce some strange bedfellows. Given the interrelatedness of China and the United States, it's possible the two countries will some day jointly fight terrorists and share military technologies. Chinese companies such as Huawei are building American communications networks—infrastructure once deemed so vital to national security that DARPA was tasked with its development—so we're already halfway there.

Indeed, the next century of technological development may be led by China. In 2008 the country for the first time became the number-two military spender in the world (France and Britain were third and fourth, respectively). The country is modernizing its arsenal, which consists largely of fifties-era Soviet weapons, but experts don't believe this is a sign that China is preparing for war; it's simply playing catch-up.[11]

China is also the biggest market in the world with a ban on pornography. The communist government is iron-fisted in its attempts to stamp out porn, to the point where it seems as though officials are more afraid of sex as a subversive force than religion or ethnicity. There's no reason to believe, however, that Chinese demand for porn is any less than it is everywhere else in the world, so its gradual creep into the country seems inevitable— and China will some day have its own sexual revolution. Perhaps the government is afraid that when the people have sexual liberty, they will demand political freedom as well.

As for food, the country is only now experiencing the kind of processing advances the United States had in the fifties. Not surprisingly, fast food is booming. Overall, China has seen double-digit growth in its food industry year on year for nearly two decades.[12] With more than a billion people to feed, there is no end to that growth in sight. It's also reasonable to expect that during this massive transformation, Chinese scientists will put the same effort into developing new food technologies as their American counterparts have. Fifty years from now we may all be eating artificial chow mein created in *Star Trek*–like replicator machines.

China is only the biggest and most obvious example of what we can expect from developing countries. Much of the rest of the world is also modernizing and pursuing new technologies. And these new technologies won't just change the stuff we have, they'll also alter how this stuff affects us, how we see the world and how we relate to each other. Technology isn't about nifty new gadgets, it's about bettering our lives. For the most part, despite what Luddites and anti-technologists may think, our lives have improved. Technology has brought us out of the dark ages by fulfilling our needs and desires. We can power our homes, feed our families, travel anywhere we want, learn about anything, answer questions, communicate with each other and acquire pretty much any object or experience we want, pretty much instantaneously. That sounds like progress to me.

NOTES

Introduction: A Shameful Trinity

[1] Huxley, Aldous, *Ends and Means: An Enquiry into the Nature of Ideals and into the Methods Employed for Their Realisation,* London, Chatto & Windus, 1937, p. 268. Copyright © 1938 by Aldous Huxley. Reprinted by permission of Georges Borchardt, Inc., for the Estate of Aldous Huxley.

[2] Author's interview with Colonel James Braden, April 2008.

[3] Author's interview with Joe Dyer, April 2009.

[4] Author's interview with Vint Cerf, March 2009.

[5] Singer, P.W., *Wired for War: The Robotics Revolution and Conflict in the Twenty-first Century,* New York, Penguin Press, 2009, p. 140.

[6] SIPRI, "World military expenditure increases despite financial crisis," June 2, 2010, www.sipri.org/media/pressreleases/2010/100602yearbook launch.

[7] From SIPRI's 2010 annual *Yearbook on Armaments, Disarmaments and National Security.*

[8] The Associated Press, "Global arms spending rises despite economic woes," June 9, 2009, www.independent.co.uk/news/world/politics/global-arms-spending-rises-despite-economic-woes-1700283.html.

[9] *Wired,* "Pentagon's Black Budget Grows to More Than $50 Billion," May 7, 2009, www.wired.com/dangerroom/2009/05/pentagons-black-budget-grows-to-more-than-50-billion.

[10] Singer, *Wired for War,* p. 140.

[11] Ibid., p. 247.

[12] Ibid., p. 239.

[13]Author's interview with John Hanke, Feb. 2009.

[14]Author's interview with Ed Zywicz, Feb. 2009.

[15]Salon.com, "The Great Depression: The Sequel," April 2, 2008, www .salon.com/opinion/feature/2008/04/02/depression.

[16]According to CostOfWar.com, a website run by the National Priorities Project.

[17]Author's interview with Brad Casemore, March 2009.

[18]Author's interview with Michael Klein, Jan. 2009.

[19]Author's interview with Evan Seinfeld and Tera Patrick, March 2009.

[20]Author's interview with Stoya, March 2009.

[21]Author's interview with Scott Coffman, March 2009.

[22]Author's interview with Ali Joone, Jan. 2009.

[23]Author's interview with Paul Benoit, March 2009.

[24]Statistics come from three places: *Los Angeles Business Journal*, "Family guy: Steve Hirsch followed in his dad's footsteps by launching his own adult film company, now the leader in a very mainstream business," Nov. 12, 2007, www .accessmylibrary.com/coms2/summary_0286-33572982_ITM; *Computerworld*, "Porn industry may decide battle between Blu-ray, HD-DVD," May 2, 2006, www.computerworld.com/s/article/print/111087/Porn_industry_ may_decide_battle_between_Blu_ray_HD_DVD_; Top Ten Reviews, "Internet Pornography Statistics," http://Internet-filter-review.toptenreviews .com/Internet-pornography-statistics.html#corporate_profiles.

[25]Author's interview with Jonathan Coopersmith, March 2009.

[26]Statistics come from a comprehensive study of the 2006 pornography market performed by the Top Ten Reviews website at http://Internet-filter-review.toptenreviews.com/Internet-pornography-statistics.html.

[27]Levenstein, Harvey, *Paradox of Plenty: A History of Social Eating in Modern America*, Berkeley, University of California Press, 2003, pp. 89–90.

[28]Ibid., p. 96.

[29]Ibid.

[30]Index Mundi, "Food and Live Animals Exports by Country in US Dollars," www.indexmundi.com/trade/exports/?section=0.

[31]Singer, *Wired for War*, p. 283.

[32]Forbes, "The World's Biggest Industry," Nov. 15, 2007, www.forbes. com/2007/11/11/growth-agriculture-business-forbeslife-food07-cx_ sm_1113bigfood.html.

[33]Author's interview with Patrick Dunne, April 2009.

[34]Weasel, Lisa, *Food Fray: Inside the Controversy Over Genetically Modified Food*, New York, Amacom, 2008, pp. 48–49.

[35]Ibid.

[36]Author's interview with Dave Rogers, Nov. 2008.

[37]Singer, *Wired for War*, p. 285.

1 **Weapons of Mass Consumption**

[1]Coventry City Council, www.coventry.gov.uk.

[2]From the Blitz Experience Museum in Coventry.

[3]*The New York Times*, "'Revenge' by Nazis," Nov. 16, 1940.

[4]*The New York Times*, "Bombs on Coventry," Nov. 16, 1940.

[5]Rhodes, Richard, *The Making of the Atomic Bomb*, New York, Simon & Schuster, 1995, p. 336.

[6]Earls, Alan R. and Edwards, Robert E., *Raytheon Company: The First Sixty Years*, Great Britain, Arcadia Publishing, 2005, p. 9.

[7]Ibid., p. 336.

[8]Ibid., p. 338

[9]Author's interview with Norman Krim, May 2008.

[10]*Reader's Digest*, "Percy Spencer and His Itch to Know," Aug. 1958, p. 114.

[11]Ibid.

[12]Earls and Edwards, *Raytheon Company*, p. 21.

[13]Norman Krim interview.

[14]Earls and Edwards, *Raytheon Company*, p. 21

[15]According to Norman Krim.

[16]Rhodes, *The Making of the Atomic Bomb*, p. 343.

[17]According to Norman Krim.

[18]Bush, Vannevar, *Science: The Endless Frontier*, New York, Arno Press, 1980, p. 330.

[19]Radiation Effects Research Foundation frequently asked questions, www.rerf.or.jp/general/qa_e/qa1.html.

[20]*The New York Times*, "Radar," May 23, 1943.

[21]Budget of the United States Government Fiscal 2009, Historical Tables, www.whitehouse.gov/omb/budget/fy2009/pdf/hist.pdf, p. 47.

[22]*The New York Times*, "Raytheon MFG: Year's Earnings Off Sharply," July 13, 1956.

[23]*The New York Times*, "Raytheon Shows Increased Profit," Aug. 29, 1945.

[24]Author's interview with Norman Krim.

[25]*The New York Times*, "War Instrument Peace-Time Boon," May 18, 1952.

[26]According to Norman Krim.

[27]Inflation calculator at www.westegg.com/inflation.

[28]Ibid.

[29]*The New York Times*, "Cooking on No Burners," May 5, 1957.

[30]*The New York Times*, "Microwaves Fail to Lure Home Cooks," Mar. 31, 1962.

[31]Ibid.

[32]*The New York Times*, "Microwave Sales Sizzle as the Scare Fades," May 2, 1976.

[33]Ibid.

[34]*The New York Times*, "Better Homes and Gadgets," Aug. 7, 2006.

[35]UK National Statistics, *Living in Britain: General Household Survey*, 2004.

[36]*The Scotsman*, "Microwaves in dire straits as iconic household purchase status is lost," Mar. 18, 2008.

[37]DuPont corporate history, http://heritage.dupont.com.

[38]Rhodes, *The Making of the Atomic Bomb*, pp. 431–32.

[39]Ibid.

[40]Ibid., p. 603.

[41]Tefal corporate history, www.tefal.co.uk/tefal/about/history.asp.

[42]*The New York Times*, "Dow Explains Sales Rise," May 30, 1950.

[43]*The New York Times*, "New Uses Shown for Saran Fiber," Sept. 17, 1952.

[44]*The New York Times*, "Peak Net Posted by Dow Chemical," Aug. 14, 1959.

[45]Fenichell, Stephen, *Plastic: The Making of a Synthetic Century*, New York, HarperCollins, 1996, p. 232.

[46]Ibid.

[47]Ibid., p. 152.

[48]Ibid., p. 183.

[49]Ibid., p. 184.

[50]I.G. Farben trial documents, www.profit-over-life.org/rolls.php?roll=97&pageID=1&expand=no.

[51]Fenichell, *Plastic*, p. 314.

[52]From testimony given in the McLibel trial, found at www.ejnet.org/plastics/polystyrene/mclibel_p6.html.

[53]*The New York Times*, "Packaging and Public Image: McDonald's Fills a Big Order," Nov. 2, 1990.

2 Better Eating Through Chemistry
Opening quotation from Roald Dahl's *Charlie and the Chocolate Factory* reprinted by permission of Puffin Books.

[1]Armstrong, Dan and Black, Dustin, *The Book of Spam*, New York, Atria Books, 2007, p. 136.

[2]Ibid.

[3]Ibid., p. 140.

[4]Ibid., p. 110.

[5]Ibid., p. 70.

[6]Levenstein, *Paradox of Plenty*, p. 85.

[7]Armstrong and Black, *The Book of Spam*, p. 73.

[8]*Cleveland Plain Dealer*, "Soldiering on: Spam and World War II," Sept. 14, 2007.

[9]Armstrong and Black, *The Book of Spam*, p. 101.

[10]*The Daily Telegraph*, "Spam at heart of South Pacific obesity crisis," Feb. 17, 2008.

[11]The Cambridge World History of Food, 2000, www.cambridge.org/us/books/kiple/potatoes.htm.

[12]Funding Universe, www.fundinguniverse.com/company-histories/The-Minute-Maid-Company-Company-History.html.

[13]Ibid.

[14]*The New York Times*, "Frozen Food Sales Growing Steadily," Mar. 9, 1950.

[15]Schlosser, Eric, *Fast Food Nation: The Dark Side of the All-American Meal*, New York, Houghton Mifflin Company, 2001, p. 114.

[16]Love, John F., *McDonald's: Behind the Golden Arches*, New York, Bantam Books, 1995, p. 329.

[17]Ibid., p. 330.

[18]Schlosser, *Fast Food Nation*, p. 115.

[19]G.I. Jobs., www.gijobs.net/magazine.cfm?issueId=61&id=809.

[20]American Frozen Food Institute, www.answers.com/topic/frozen-fruits-fruit-juices-vegetables.

[21]Market Line Frozen Food Global Industry Guide, www.globalbusinessinsights.com/content/ohec0101m.pdf.

[22]World Health Organization, www.who.int/topics/obesity/en.

[23]The Daily Plate, www.thedailyplate.com/nutrition-calories/food/generic/french-fries.

[24]*Journal of Dairy Science*, 1956, pp. 844–45.

[25]Ibid.

[26]Ibid.

[27]Ibid.

[28]Author's interview with Howard Rogers, Aug. 2008.

[29]WWII Navy Food Remembered, www.foodhistory.com/foodnotes/leftovers/ww2/usn/pla.

[30]Infoshop report on eggs, www.the-infoshop.com/study/mt42271-us-eggs.html.

[31]Leffingwell and Associates report, www.leffingwell.com/top_10.htm.

[32]Author's interview with Scott Smith, June 2009.

[33]Levenstein, *Paradox of Plenty*, p. 109.

[34]Ibid., p. 112.

[35]Ibid., p. 21.

[36]Ibid., p. 22.

[37]Ibid.

[38]Ibid., p. 69.

[39]Connor, John M. & Schiek, William A., *Food Processing: An Industrial Powerhouse in Transition,* Second Edition, New York, John Wiley & Sons, 1997, p. 33.

[40]Love, *McDonald's: Behind the Golden Arches*, p. 16.

[41]Ibid., p. 121.

[42]Ibid., p. 334.

[43]Company website, www.keystonefoods.com.

[44]Love, *McDonald's: Behind the Golden Arches*, p. 342.

[45]Ibid., p. 342.

[46]Subway surpassed McDonald's in terms of total outlets in 2011 but lags far behind in revenue.

[47]Ibid., p. 347.

3 Arming the Amateurs

[1]Di Lauro, Al and Rabkin, Gerald, *Dirty Movies: An Illustrated History of the Stag Film 1915–1970*, New York, Chelsea House Publishers, 1975, p. 52.

[2]*Playboy*, "The History of Sex in Cinema: The Stag Film," Nov. 1976.

[3]Di Lauro and Rabkin, *Dirty Movies*, p. 55.

[4]Ibid., pp. 53–54.

[5]*Playboy*, "The History of Sex in Cinema."

[6]Di Lauro and Rabkin, *Dirty Movies*, p. 59.

[7]Zimmerman, Patricia R., *Reel Families: A Social History of Amateur Film*, Bloomington, Indiana University Press, 1995, p. 96.

[8]Editors of *Look, Movie Lot to Beachhead: The Motion Picture Goes to War and Prepares for the Future*, Garden City, N.Y., Doubleday, 1945, pp. 58–59.

[9]Carl Preyer, "Movies Report on Defense Programs," *American Cinematographer*, Aug. 1943, p. 445.

[10]C.E. Eraser, "Motion Pictures in the United States Navy," *Journal of the Society of Motion Picture Engineers*, Dec. 1932, pp. 546–52.

[11]Zimmerman, *Reel Families*, p. 97.

[12]*The New York Times*, "Fastest Movie Camera Scans the War Machine," Aug. 15, 1943.

[13]Zimmerman, *Reel Families*, p. 108.

[14]*American Cinematographer*, "A.S.C. and the Academy to Train Cameramen for Army Services," June 1942, p. 255.

[15]Thompson, George Raynor, "Overview: The Signal Corps in World War II," *Army Communicator, Special Edition: The Signal Corps in World War II*, vol. 20, no. 4.

[16]Zimmerman, *Reel Families*, pp. 115–16.

[17]Ibid., pp. 120–21.

[18]Coopersmith, Jonathan, "Pornography, Technology and Progress," in *Icon: Journal of the International Committee for the History of Technology*, volume 4, 1998, p. 101.

[19]Lane, Frederick S., III, *Obscene Profits: The Entrepreneurs of Pornography in the Cyber Age*, New York, Routledge, 2001, p. 47.

[20]U.S. Commission on Obscenity and Pornography, *The Report of the Commission on Obscenity and Pornography*, Washington D.C., GPO, 1970, pp. 129, 190.

[21]Watts, Steven, *Mr. Playboy: Hugh Hefner and the American Dream*, Hoboken, N.J., John Wiley & Sons Inc., 2008, p. 59.

[22]Ibid., p. 70.

[23]McDonough, Jimmy, *Big Bosoms and Square Jaws: The Biography of Russ Meyer, King of the Sex Film*, New York, Crown, 2005, p. 125.

[24]Ibid., p. 108.

[25] Ibid., p. 109.

[26] Ibid., p. 4.

[27] Lane, *Obscene Profits*, p. 29.

[28] McDonough, *Big Bosoms and Square Jaws*, p. 99.

[29] Ibid., p. 210.

[30] Lane, *Obscene Profits*, p. 29.

[31] From Lasse Braun's official website, www.lassebraun.com.

[32] Lane, *Obscene Profits*, p. 48.

[33] From the American Presidency Project, www.presidency.ucsb.edu/ws /index.php?pid=2759.

[34] Coopersmith, Jonathan, "Do-It-Yourself Pornography," *Ars Electronika*, 2008.

[35] Author's interview with Gary Cole, May 2009.

[36] Coopersmith, "Do-It-Yourself Pornography."

[37] *Chicago Tribune*, "The strange allure of photo booths," May 25, 2005, www .chicagotribune.com/topic/chi-040512photostory,0,6382269.story.

[38] Coopersmith, "Do-It-Yourself Pornography."

[39] Author's interview with Jonathan Coopersmith, March 2009.

[40] Coopersmith, "Do-It-Yourself Pornography."

[41] Author's interview with Brad Casemore, March 2009.

4   A Game of War

[1] *The New York Times*, "G.I. Joe Doll Is Capturing New Market," July 24, 1965.

[2] Slinky history, www.poof-slinky.com/Slinky-Museum/Slinky-History.

[3] *The New York Times*, "Talking Toys with Betty James," Feb. 21, 1996.

[4] Slinky history.

[5] Silly Putty timeline, www.sillyputty.com.

[6] Ibid.

[7] *The New Yorker*, "Talk of the Town," Aug. 25, 1950.

[8] Silly Putty timeline.

[9] Massachusetts Institute of Technology Inventor of the Week Archive, http://web.mit.edu/invent/iow/sillyputty.html.

[10] Lord, M.G., *Forever Barbie: The Unauthorized Biography of a Real Doll*, New York, William Morrow, 1994, p. 25.

[11] Ibid., p. 26.

[12] Ibid., p. 27.

[13] Ibid., p. 20.

[14] Ibid., p. 24.

[15] Author's interview with Norman Krim.

[16] *People*, "Jack Ryan and Zsa Zsa: A Millionaire Inventor and His Hungarian Barbie Doll," July 14, 1975.

[17] Gabor, Zsa Zsa, *One Lifetime Is Not Enough*, New York, Delacorte Press, 1991, p. 235.

[18] Lord, *Forever Barbie*, p. 30.

[19] Ibid., p. 32.

[20] BBC News, "Vintage Barbie struts her stuff," Sept. 22, 2006.

[21] Hot Wheels history, www.swflorida-hwc.com/page_14.htm.

[22] Author's interview with Norman Krim.

[23] *Creative Computing Video & Arcade Games*, "Who really invented the video game?" Vol. 1, No. 1, Spring 1983.

[24] "The Manhattan Project, an Interactive History," www.mbe.doe.gov /me70/manhattan/trinity.htm.

[25] *The New York Times*, "There Is No Defense Against Atomic Bombs," Nov. 3, 1946.

[26] *Creative Computing Video & Arcade Games*, "Who really invented the video game?"

[27] Author's interview with Dee Katz, Sept. 2008.

[28] Author's interview with Robert Dvorak Jr., Sept. 2008.

[29] *Creative Computing Video & Arcade Games*, "Who really invented the video game?"

[30] Author's interviews with Ralph Baer, Oct. 2008.

[31] U.S. patent number 3,728,480.

[32] PBS biography, www.pbs.org/transistor/album1/shockley/index.html.

[33] "The Time 100," www.time.com/time/time100/scientist/profile /shockley.html.

[34] *Computerworld*, "The Transistor: The 20th Century's Most Important Invention," Jan. 3, 2008, www.arnnet.com.au/article/202801 /transistor_20th_century_most_important_invention.

[35] William Shockley, Nobel lecture 1956, see http://nobelprize.org/nobel_ prizes/physics/laureates/1956/shockley-lecture.html.

[36] *The New York Times*, "G.I. Joe Doll Is Capturing New Market," July 24, 1965.

37 Seeking Alpha, "The video game industry: An 18-billion entertainment juggernaut," Aug. 5, 2008, http://seekingalpha.com/article/89124-the-video-game-industry-an-18-billion-entertainment-juggernaut.

38 CBC News, "U.S. Army forms unit to enlist video games," Nov. 24, 2008, www.cbc.ca/technology/story/2008/11/24/tech-army.html?ref=rss.

39 Singer, *Wired for War*, p. 365.

40 Ibid., p. 395.

5   Food from the Heavens

1 From Plutarch's *Parallel Lives.*

2 Annual total comes from *The New York Times*, "Starship Kimchi: A Bold Taste Goes Where It Has Never Gone Before," Feb. 24, 2008, www.nytimes.com/2008/02/24/world/asia/24kimchi.html. I've divided that consumption by the number of households in 2008, about 16.7 million, from The Hankyoreh, "Household debt soars to record high of 660 trillion won," Sept. 5, 2008, http://english.hani.co.kr/arti/english_edition/e_business/308655.html.

3 *The New York Times*, "Starship Kimchi."

4 Ibid.

5 Ibid.

6 Allied casualties come from Cooksley, Peter G, *Flying Bomb*, New York, Charles Scribner's Sons, 1979, p. 175. Concentration camp numbers are from Global Security's history of weapons of mass destruction, www.globalsecurity.org/wmd/ops/peenemunde.htm.

7 Nine countries have developed nuclear weapons: the United States, Russia, the United Kingdom, France, China, India, Pakistan, Israel and North Korea. Iran is a possible tenth. Twelve countries have launched rockets into space: all of the nuclear powers except Pakistan and North Korea, but including Iran, plus Canada, Australia, Japan and Ukraine.

8 Glenn, John, *John Glenn: A Memoir*, New York, Bantam Books, 1999, p. 264.

9 My visit took place on April 1, 2009.

10 I have to admit the beans were better at Goode Company.

11 White, Terry, *The SAS Fighting Techniques Handbook*, Guilford, Connecticut, Globe Pequot Press, 2007, p. 28.

[12] According to Natick's senior advisor in nutritional biochemistry, Dr. Patrick Dunne, interviewed in March 2009.

[13] NASA website, www.nasa.gov/offices/ogc/about/space_act1.html# FUNCTIONS.

[14] Examples taken from NASA's *Spinoff*, 2008.

[15] NASA, "Thought for Food," *Spinoff*, 1977, p. 85.

[16] NASA, "Food Service System," *Spinoff*, 1992, pp. 78–79.

[17] I love camping, but I have yet to find decent-tasting camping food.

[18] NASA, "Tenderness Tester," *Spinoff*, 1977, p. 86.

[19] NASA, "Poultry Plant Noise Control," *Spinoff*, 1982, pp. 94–95.

[20] NASA, "Spinoff from Space Fuel," *Spinoff*, 1982, pp. 46–49.

[21] Officially known as "Poppin' Fresh," the doughboy wasn't actually conceived until 1965.

[22] NASA, "A Dividend in Food Safety," *Spinoff*, 1991, pp. 52–53.

[23] Ibid., pp. 53–54.

[24] NASA, "Space Research Fortifies Nutrition Worldwide," *Spinoff*, 2008, pp. 106–7.

[25] NASA, "Eating on Demand," *Spinoff*, 1998, p. 73, and an Enersyst report at http://ift.confex.com/ift/2002/techprogram/paper_10060. htm. Cooking times taken from "Enersyst's Speed Cooling, Thawing Technology Available for Home Use," *Appliance Design*, Aug. 22, 2001, www.appliancedesign.com/Articles/Breaking_News /8731be2d0b938010VgnVCM100000f932a8c0____.

[26] *The National Provisioner*, "Retort pouches heating up," May 1, 2005, www .allbusiness.com/manufacturing/food-manufacturing/460996-1.html.

[27] Author's interview with Dave Williams, March 2009.

[28] Harvey, Brian, *The Rebirth of the Russian Space Program: 50 Years After Sputnik, New Frontiers*, Chichester, UK, Praxis Publishing, 2007, p. 80.

[29] Ibid., p. 284.

[30] Ibid., p. 283.

[31] Space.com, "Russia opens space tech transfer office," July 6, 2000, www.space.com/businesstechnology/technology/russian_ technology_000706.html.

[32] Cohon, George, with Macfarlane, David, *To Russia with Fries*, Toronto, McClelland & Stewart Inc., 1997, p. 176.

[33] Love, *Behind the Golden Arches*, p. 439.

³⁴Cohon and Macfarlane, *To Russia with Fries*, p. 178.

³⁵Author's interview with Shoichi Tachibana, April 2009.

³⁶Author's interview with Amrinder Singh Bawa, April 2009.

³⁷*The Daily Telegraph*, "Giant space vegetables 'could feed the world,'" May 12, 2008, www.telegraph.co.uk/news/uknews/1949129/Giant-space-vegetables-could-feed-the-world.html.

³⁸*The Hindu Business Line*, "McDonald's growth in India hit by poor infrastructure," Aug. 15, 2004, www.thehindubusinessline.com/bline/2004/08/16/stories/2004081600510500.htm. The 2009 figures were taken from the respective corporate websites.

6   The Naked Eye Goes Electronic

¹Flynt, Larry, *Sex, Lies and Politics: The Naked Truth*, New York, Kensington Books, 2004. Reprinted by arrangement with Kensington Publishing Corp., www.kensingtonbooks.com. All rights reserved.

²Author's interview with Lena Sjööblom, Feb. 2008.

³Watts, *Mr. Playboy*, p. 209.

⁴*Playboy*, "Swedish Accent," Nov. 1972.

⁵Signal & Imaging Processing Institute history, http://sipi.usc.edu.

⁶From an estimate made by Jeff Seideman during an interview in Feb. 2008.

⁷Author's interview with Alexander Sawchuk, Feb. 2008.

⁸According to W. Scott Johnston, now a program manager for Raytheon. Johnston originally scanned the photo.

⁹*IEEE Personal Communication Society Newsletter*, "Culture, communication and an information age Madonna," May/June 2001.

¹⁰Author's interview with Jeff Seideman, Feb. 2008.

¹¹*Optical Engineering*, "Editorial," Jan. 1992.

¹²Author's interview with David Munson, Feb. 2008.

¹³Author's interview with Kevin Craig, May 2009.

¹⁴Details taken of the IS&T convention appearance are taken from *The Journal of the Photographic Historical Society of New England*, issue 1, 1999, and the interview with Jeff Seideman.

¹⁵Eastman Kodak corporate history, www.kodak.com.

¹⁶PluggedIn Kodak corporate blog, "We had no idea," Oct. 16, 2007, http://stevesasson.pluggedin.kodak.com/default.asp?item=687843.

[17] Autor's interview with Steven Sasson, Dec. 2009.

[18] The Worldwide Community of Imaging Associations, 2001–2002 PMA Industry Trends Report International Markets.

[19] CNET News, "Cheaper DLSRs to drive digital camera sales," Aug. 13, 2007, http://news.cnet.com/8301-10784_3-9759018-7.html.

[20] National Reconnaissance Project, www.nro.gov/corona/facts.html.

[21] Kroc, Ray, *Grinding It Out: The Making of McDonald's*, New York, St. Martin's Paperbacks, 1987, p. 6.

[22] Schlosser, *Fast Food Nation*, p. 66.

[23] NASA's Land Remote Sensing Policy, http://geo.arc.nasa.gov/sge /landsat/15USCch82.html.

[24] Author's interview with John Hanke, Feb. 2009.

[25] *The New York Times*, "The Stock? Whatever. Google Keeps on Innovating," Oct. 31, 2004.

[26] *The New York Times*, "Google Offers a Bird's-Eye View, and Some Governments Tremble," Dec. 20, 2005.

[27] *Broadcasting and Cable Yearbook 1995*, New Providence, New Jersey, 1995, p. xxi.

[28] *The Economist*, "An Adult Affair," Jan. 4, 1997.

[29] *Los Angeles Times*, "Porn quietly becoming pay-TV gold mine," July 6, 2000.

[30] *The New York Times*, "Videotapes for Homes," June 13, 1979.

[31] Ibid.

[32] *The New York Times*, "Sex Films Find Big Market in Home Video," April 5, 1979.

[33] *Icon: Journal of the International Committee for the History of Technology*, volume 4, 1998, p. 105.

[34] *The New York Times*, "Sex Films Find Big Market in Home Video," April 5, 1979.

[35] Lane, *Obscene Profits*, p. 51.

[36] *The New York Times*, "Company News: Sony, in a Shift, to Make VHS Units," Jan. 12, 1988.

[37] *The New York Times*, "U.S. Move in Video Cameras," Jan. 5, 1984.

[38] *Forbes*, Sept. 18, 1978.

[39] From a Nov. 29, 2002, radio transcript on www.onthemedia.org.

[40] *Video Review*, March 1984.

[41]*The New York Times*, "Camcorder, CD Sales May Double in 1986," June 2, 1986.

[42]Department of Justice, *Attorney General's Commission on Pornography: Final Report*, two volumes, Washington, 1986.

[43]Ibid.

[44]*New Society*, Sept. 18, 1987.

[45]*The New York Times*, "Justices Uphold Businesses' Right to Sell Phone Sex," June 24, 1989.

[46]*The Economist*, "Heavy breathing," July 30, 1994.

[47]*The Washington Post*, "Money flows into poor countries on X-rated phone lines," Sept. 23, 1996.

[48]Family Safe Media pornography statistics, www.familysafemedia.com/pornography_statistics.

[49]*China Daily*, "Ninety-seven phone sex lines closed in China," Jan. 8, 2005.

[50]Adult Video News, "Report: Adult mobile market still growing," May 6, 2009, http://business.avn.com/articles/35226.html.

[51]Author's interview with Michael Klein, Jan. 2009.

[52]Author's interview with Kim Kysar, March 2009.

7 The Internet: Military Made, Porn Perfected

[1]*The Dallas Morning News*, "Unlikely innovators: Many online technologies were first perfected by the adult industry," April 26, 2001.

[2]Author's interview with Tera Patrick, March 2009.

[3]Friedman, Thomas L., *The World Is Flat*, New York, Farrar, Straus & Giroux, 2005.

[4]Hafner, Katie, *Where Wizards Stay Up Late: The Origins of the Internet*, New York, Simon & Schuster, 1998, p. 20.

[5]From a speech at DARPAtech by Tony Tether on Aug. 7, 2007.

[6]From a report given by Tony Tether to the House of Representatives' subcommittee on terrorism, Mar. 13, 2008.

[7]Author's interview with Vint Cerf, Feb. 2009.

[8]He was also officially made a vice-president.

[9]Bidgoli, Hossein, *The Internet Encyclopedia*, Hoboken, New Jersey, Wiley & Sons, 2001, p. 118.

[10]*Icon: Journal of the International Committee for the History of Technology*, volume 4, 1998, p. 110.

[11]*The Dallas Morning News*, "Unlikely innovators," April 26, 2001.

[12]*The Daily Telegraph*, "Three loud cheers for the father of the web," Jan. 28, 2005.

[13]*Publishing Executive*, "Integrated publishing ad sales strategies that work," Feb. 9, 2007.

[14]Coopersmith, "Pornography, Technology and Progress," p. 110.

[15]Ibid., p. 111.

[16]*The New York Times*, "Digital 'watermarks' assert Internet copyright," June 30, 1997.

[17]CNET News, "Playboy wins Net copyright suit," April 2, 1998, http://news.cnet.com/Playboy-wins-Net-copyright-suit/2100-1023_3-209775.html.

[18]Coopersmith, "Pornography, Technology and Progress," p. 113.

[19]Author's interview with Steve Orenstein, March 2009.

[20]The interview took place in April 2009.

[21]Author's interview with Paul Benoit, March 2009.

[22]Author's interview with Jack Dowland, April 2009.

[23]ZDNet, "Naughton looks set to escape prison," http://news.zdnet.co.uk/emergingtech/0,1000000183,2080719,00.htm.

[24]Techcrunch, "Internet pornography stats," May 12, 2007, www.techcrunch.com/2007/05/12/Internet-pornography-stats.

[25]Coopersmith, "Pornography, Technology and Progress," p. 113.

[26]Ibid., p. 112.

[27]*The Independent*, "Danni Ashe: Danni's drive to Net profits," Aug. 6, 2001.

[28]*The Dallas Morning News*, "Unlikely innovators," April 26, 2001.

[29]Ibid.

[30]VUNet.com, "ISPs get an eyeful of porn bonanza," Sept. 10, 2001.

[31]BBC News, "Porn and music drive broadband," May 20, 2003, http://news.bbc.co.uk/2/hi/technology/2947966.stm.

[32]VUNet.com, "ISPs get an eyeful of porn bonanza."

[33]The $3 billion amount comes from a TechCrunch estimate for 2006, www.techcrunch.com/2007/05/12/Internet-pornography-stats. The $5 billion global number comes from the Internet Filter Review, http://Internet-filter-review.toptenreviews.com/Internet-pornography-statistics.html.

[34]*Adult Video News*, "Analysis: Porn cheaper than ever online," April 21, 2009, http://business.avn.com/articles/printable/35077.html.

[35]Techcrunch, "Internet pornography stats," May 12, 2007, www.techcrunch.com/2007/05/12/Internet-pornography-stats.

[36]*Adult Video News*, "Analysis: Porn cheaper than ever online."

[37]*Wired*, "Turns out porn isn't recession proof," http://blog.wired.com/business/2008/07/turns-out-por-1.html.

[38]Author's interview with Samantha Lewis, April 2009.

[39]Author's interview with Kim Kysar, March 2009,

[40]Author's interview with Michael Klein, Jan. 2009.

[41]Boston.com, Nov. 5, 2006, "Miles away, 'I'll have a burger,'" www.boston.com/business/globe/articles/2006/11/05/miles_away_ill_have_a_burger.

[42]Customer Management Insight, "McDonald's, cell centers and Joe Fleischer," April 11, 2006, www.callcentermagazine.com/blog/archives/2006/04/mcdonalds_call.html.

8   Seeds of Conflict

[1]From his 1949 Nobel lecture, http://nobelprize.org/nobel_prizes/peace/laureates/1949/orr-lecture.html.

[2]CNN.com, "President George W. Bush's address on stem cell research," Aug. 11, 2001, http://edition.cnn.com/2001/ALLPOLITICS/08/09/bush.transcript/index.html.

[3]Wilmut, Ian and Highfield, Roger, *After Dolly: The Promise and Perils of Human Cloning*, New York, W.W. Norton & Company Ltd., 2006, p. 186.

[4]MSNBC.com, "Bush administration in hot seat over warming," Jan. 30, 2007, www.msnbc.msn.com/id/16886008.

[5]*The New York Times*, "The science of denial," June 4, 2008.

[6]United Nations wire, "U.S. mulls action against Europe," Jan. 10, 2003, www.unwire.org/unwire/20030110/31347_story.asp.

[7]*The Independent*, "George Bush: Europe must accept GM food," May 23, 2003, www.independent.co.uk/opinion/commentators/george-bush-europe-must-accept-gm-food-590957.html.

[8]Weasel, *Food Fray*, p. 61.

[9]Ibid., pp. 61–62.

[10]Ibid., p. 62.

[11]Nobel lecture by Norman Borlaug, "The Green Revolution, peace and humanity," Dec. 11, 1970, http://nobelprize.org/nobel_prizes/peace/laureates/1970/borlaug-lecture.html.

[12]National Creutzfeldt-Jakob Disease Surveillance Unit, www.cjd.ed.ac.uk/index.htm.

[13]*Daily Mail*, "Questions about genetically modified organisms," June 1, 1999.

[14]*The Wall Street Journal*, "McDonald's, other fast-food chains pull Monsanto's bio-engineered potato," April 28, 2000.

[15]Reuters, "Eli Lilly and Company to acquire Monsanto's Posilac brand dairy product and related business," Aug. 20, 2008, www.reuters.com/article/pressRelease/idUS127863+20-Aug-2008+PRN20080820.

[16]Weasel, *Food Fray*, p. 6.

[17]Statistics are from the International Service for the Acquisition of Agri-Biotech Applications Global Status of Commercialized Biotech/GM Crops: 2008.

[18]Center for Food Safety statistics.

[19]Reuters, "Monsanto sues Germany over GMO maize ban," April 21, 2009, www.reuters.com/article/marketsNews/idUSLL62523620090421.

[20]*The Daily Telegraph*, "Prince Charles warns GM crops risk causing the biggest-ever environmental disaster," Nov. 10, 2008.

[21]Singer, *Wired for War*, p. 283.

[22]Singer, P.W., *Children at War*, Los Angeles, University of California Press, 2006, p. 39.

[23]A 2009 estimate by the Institute of Food Technologists.

[24]*Wired*, "The future of food," Nov. 2008.

[25]Living History Farm, "The Green Revolution," www.livinghistoryfarm.org/farminginthe50s/crops_13.html.

[26]Woodward, Billy, Shurkin, Joel and Gordon, Debra, *Scientists Greater Than Einstein: The Biggest Lifesavers of the Twentieth Century*, Fresno, California, Linden Publishing, 2009, chapter 5.

[27]From the Monsanto website, www.monsanto.com.

[28]Author's interview with Peter Singer, May 2009.

[29]Singer, *Children at War*, pp. 39–40.

[30]Ibid., p. 45.

[31]Ibid., p. 63.

[32] Ibid.

[33] Author's interview with Graeme Smith, May 2009.

[34] *USA Today*, "Root out seeds of terrorism in sub-Saharan countries," April 14, 2003, www.usatoday.com/news/opinion/columnist /wickham/2003-04-14-wickham_x.htm.

[35] *The New York Times*, "War in the Gulf: The troops—blacks wary of their big role as troops," Jan. 25, 1991.

[36] *The New York Times*, "'Counter-recruiter' seeks to block students' data from the military," Oct. 23, 2008.

[37] Author's interview with Eric Darier, May 2009.

[38] Weasel, *Food Fray*, p. 68.

[39] Ibid., p. 69.

[40] Author's interview with Bruce Cranny, May 2009.

[41] Weasel, *Food Fray*, p. 79.

9   Fully Functional Robots

[1] *The Economist*, "Trust me, I'm a robot," June 8, 2006, www.economist .com/sciencetechnology/tq/displaystory.cfm?story_id=7001829.

[2] Author's interview with Bakhtiar Litkouhi, GM's manager of vehicle control systems, Jan. 2008.

[3] The 2008 estimate and $21.4 billion projection comes from a report by Electronics.ca Publications, "Global robotics market to reach $21.4 billion in 2014," April 23, 2009, at www.pcb007.com/pages/zone.cgi?a=49647. The $100 billion projection comes from Robotics TCMNet, "S. Korean gov't aims to turn robot industry into global leader," April 17, 2008, http:// robotics.tmcnet.com/news/2009/04/17/4137987.htm.

[4] *Scientific American*, "A robot in every home," Jan. 2007, www.sciam.com /article.cfm?id=a-robot-in-every-home.

[5] Comments are from a speech at the RoboBusiness conference in Boston, April 2009.

[6] Author's interview with Colin Angle, April 2008.

[7] From full-year 2008 company earnings.

[8] Author's interview with Kevin Fahey, April 2008.

[9] DARPA press release, "DARPA plans grand challenge for robotic ground vehicles," Jan. 2, 2003, www.darpa.mil/grandchallenge04 /media_news.htm.

[10] Ibid.

[11] Singer, *Wired for War*, p. 137.

[12] *The Washington Post*, "The Army's $200 billion makeover," Dec. 7, 2007, www.washingtonpost.com/wp-dyn/content/article/2007/12/06 /AR2007120602836.html?sid=ST2007120602927.

[13] CBCNews.ca, "U.S. Army praises robot makers for help in wars," April 8, 2008, www.cbc.ca/news/story/2008/04/08/tech-robo-show.html.

[14] Singer, *Wired for War*, p. 78.

[15] Author's interview with Takayuki Toriyama, April 2009.

[16] Singer, *Wired for War*, p. 241.

[17] *The Daily Telegraph*, "Cyberlover flits its way to Internet fraud," Dec. 11, 2007, www.telegraph.co.uk/news/uknews/1572077/Cyberlover-flirts-its-way-to-Internet-fraud.html.

[18] Rough Type, "Slutbot aces Turing test," Dec. 8, 2007, www.roughtype. com/archives/2007/12/slutbot_passes.php.

[19] Hacker News, Dec. 9, 2007, http://news.ycombinator.com /item?id=87426.

[20] Author's interview with Dean Turner, March 2009.

[21] It typically ranks in the high 500s on Alexa.com.

[22] Author's interview with Scott Coffman, March 2009.

[23] From the Real Doll website, www.realdoll.com. Real Doll sales figures come from Levy, David, *Love + Sex with Robots: The Evolution of Human-Robot Relationships*, New York, HarperCollins, 2007, p. 244.

[24] Levy, *Love + Sex with Robots*, p. 175.

[25] *Japan Today*, "Plastic fantastic: Japan's doll industry booming," May 19, 2008.

[26] Author's interview with Le Trung, March 2009.

[27] Levy, *Love + Sex with Robots*, p. 215.

[28] Author's interview with Douglas Hines, Dec. 2009.

[29] Author's interview with R. Craig Coulter, April 2008.

[30] CNET News, "HyperActive Bob gets fries to go," Mar. 27, 2007, http:// news.cnet.co.uk/gadgets/0,39029672,49288771,00.htm.

[31] Author's interview with David Rogers, senior director of business integration for McDonald's Canada, April 2008.

[32] *Robotics Business Review*, March/April 2009 issue, pp. 30–31.

[33]The Reuters news video can be found on my blog at www
.bombsboobsburgers.net/2009/05/pizza-making-robots-no-boogers.
html.

10 Operation Desert Lab

[1]Wilder, Thornton, *The Skin of Our Teeth*, New York, Samuel French, 1942, p.
117. Copyright © 1942 The Wilder Family LLC. Reprinted by arrangement
with The Wilder Family LLC and the Barbara Hogenson Agency.

[2]Singer, *Wired for War*, p. 310.

[3]The Associated Press, "Video shows direct hit by smart bomb," Jan. 19,
1991.

[4]*New Scientist*, "US gambles on a 'smart' war in Iraq," Mar. 19, 2003, www
.newscientist.com/article/dn3518-us-gambles-on-a-smart-war-in-
iraq.html.

[5]Global Security, "Night vision goggles," www.globalsecurity.org/military
/systems/ground/nvg.htm.

[6]*The Gazette* (Colorado Springs), "Military learns from Gulf War glitches,
updates space technology," Jan. 27, 2001, www.globalsecurity.org/org
/news/2001/010127-space2.htm.

[7]White House Office of the Press Secretary, "Statement by the President
regarding the United States' decision to stop degrading Global Position
System accuracy," May 1, 2000, http://clinton4.nara.gov/WH/EOP
/OSTP/html/0053_2.html.

[8]Garmin 2008 annual report.

[9]Marketresearch.com, "World GPS market forecast to 2013," www
.marketresearch.com/product/display.asp?productid=2215740.

[10]Singer, *Wired for War*, p. 58.

[11]Chronology of Microsoft Windows Operating Systems, www.islandnet
.com/~kpolsson/windows/win1990.htm.

[12]Singer, *Wired for War*, p. 58.

[13]CBC News, "Honda unveils wearable walking device," www.cbc.ca
/world/story/2008/11/07/walk-assist.html.

[14]Raytheon press release, www.raytheon.com/newsroom/technology
/rtn08_exoskeleton.

[15]*Wired*, "Be more than you can be," March 2007.

[16]Michael Goldblatt speech, DARPAtech 2002.

[17]*Wired*, "Be more than you can be," March 2007.

[18]Author's interview with Franz-Josef Och, Feb. 2009.

[19]*The Washington Post*, "Tongue in check," May 24, 2009, www.washington post.com/wp-dyn/content/article/2009/05/21/AR2009052104697 .html.

[20]Ibid.

[21]Google blog, "A new landmark in computer vision," http://googleblog .blogspot.com/2009/06/new-landmark-in-computer-vision.html.

[22]"Plan for cloaking device unveiled," May 25, 2006, BBC News, http:// news.bbc.co.uk/2/hi/5016068.stm.

[23]*The Sun*, "Boffins invent invisible tank," Oct. 30, 2007, www.thesun .co.uk/sol/homepage/news/article403250.ece.

Conclusion: The Benevolence of Vice

[1]Alcott, Amos Bronson, *Table Talk*, Boston, Roberts Brothers, 1877, p. 90.

[2]Author's interview with Ed Zywicz, Feb. 2009.

[3]Author's interview with George Caporaso, Feb. 2009.

[4]Mutsugoto website, www.mutsugoto.com/concept.html.

[5]Author's interview with Joanna Angel, July 2009.

[6]Author's interview with Stoya, March 2009.

[7]I nicked this subhead from the book of the same name by Harvey Levenstein.

[8]*Chris Rock: Bring the Pain*, 1996.

[9]Hoover Institution Policy Review, "Is Food the New Sex?" Feb. & March 2009, www.hoover.org/publications/policyreview/38245724.html.

[10]CBC News, "Researchers working on natural food preservatives," Sept. 7, 2009, www.cbc.ca/health/story/2009/09/07/natural-organic-food-preservatives.html.

[11]*The Independent*, "Global arms spending rises despite economic woes," June 9, 2009, www.independent.co.uk/news/world/politics/global-arms-spending-rises-despite-economic-woes-1700283.html.

[12]USDA, "China's food service sector continues sustained growth," July 2006, www.fas.usda.gov/info/fasworldwide/2006/07-2006 /ChinaHRIOverview.htm.

ACKNOWLEDGMENTS

I've wanted to write a book for nearly half my life, yet I had no idea how much work it would actually be when I finally got to it. This book simply wouldn't have happened without the help, patience and understanding of many people. First and foremost is my literary agent John Pearce, who helped me transform the seed of an out-there idea into a cogent thesis. John didn't really know what to make of my weird technology book at first, but he warmed to it quickly, and from then on there was no stopping him. While I went through the roller coaster ride that is writing—sometimes loving, sometimes hating what I'd written—John was always enthusiastic, never failing to bring me back to the right frame of mind. His temperament was shared by everyone at Westwood Creative Artists, particularly Natasha Daneman and Michael Levine, who also helped buoy the journey with occasional announcements of good news.

Of course, none of it would have been possible were it not for Nicole Pointon, an old friend from school who used to work in book publishing. Nicole pointed me in John's direction when I asked her if she knew any good agents, and needless to say, she's a very good judge of such things.

Putting the book together was a laborious and difficult task in that there were three very different industries to research. The scientists, engineers, inventors, company executives, analysts, professors, writers, journalists and even astronauts and porn stars are simply too numerous to thank individually, but I'm extremely appreciative of the time each gave to talk to me about their area of expertise, sometimes for hours.

A pair of companies warrant singling out, though, for being extraordinarily kind with their time: Google and Digital Playground. Tamara Micner, Dan Martin and Karen Wickre (in Toronto, Washington and Mountain View, California, respectively) helped immensely by making some key Googlers available to talk about all aspects of technology, from search engines to satellites, which helped inform and shape my thoughts about the themes covered in this book. I'd like to thank Vint Cerf, Rick Whitt, Matt Cutts, Franz Och, John Hanke and Dan Slater for sharing their knowledge, and for doing so in an open and often humorous manner.

At Digital Playground, Chris Ruth worked tirelessly to put me in touch with the right people and supplied me with one of my life's most interesting experiences—certainly few writers can say they've had dinner with a gaggle of porn stars. I'd like to thank Ali Joone, Samantha Lewis and Farley Cahen for taking the time to discuss their business with me on several occasions, as well as their stars Jesse Jane and Stoya, for sharing their insights. Their views helped me see things from a different perspective and were much appreciated, despite Stoya's saying I was "old" for making her call me on a landline.

I also owe a debt a gratitude to my employers at the Canadian Broadcasting Corporation, who gave me the time off I needed to work on this book. My friends and colleagues Ian Johnson and Andre Mayer each provided an invaluable edit and critique of my manuscript when it was done. Both pointed out some things that didn't work and drew attention to some of my more annoying writing habits (which I have hopefully corrected). Alex Schultz and then Helen Reeves at Penguin (Canada) also flexed their considerable editorial muscles in whipping the manuscript into shape, while Scott Steedman did an excellent job in tightening it up through his copy edit. I must thank Jo Paul, who initially oversaw the Australian side of things for Allen & Unwin, and Sue Hines for riding the manuscript through to its conclusion. I'm also grateful to James Jayo and the crew at Lyons Press for putting the finishing touches on the U.S. version. I am extremely indebted to everyone who edited the manuscript, as they helped shape the raw words I came up with into what I hope is an eminently readable book.

I must also thank my friends and family and ask their forgiveness for the neglect I was sometimes forced to visit on them. There were a number of birthday parties and social events I missed as I furiously tried to meet deadlines—I can only hope they understand, and next time I promise to organize my time better. I have to extend a special thanks to Claudette, who perhaps endured more of this than anyone. Her patience, understanding and encouragement rank her up there with the kindest of saints.

Last but not least, I must thank you, dear reader, for picking up this book. I hope you enjoy it as much as I enjoyed writing it.

And let's not forget Paris Hilton, without whom none of this would have happened.

INDEX